Renewing Cities with Value Capture Planning

Raymond Rauscher

Renewing Cities with Value Capture Planning

Model for Achieving Equitable Housing, Public and Open Spaces, and Sustainable Transport

Springer

Raymond Rauscher
Newcastle Business School
University of Newcastle
Newcastle, NSW, Australia

ISBN 978-3-030-62957-1 ISBN 978-3-030-62958-8 (eBook)
https://doi.org/10.1007/978-3-030-62958-8

This Springer imprint is published by the registered company Springer Nature Switzerland AG
The registered company address is: Gewerbestrasse 11, 6330 Cham, Switzerland

There are many people in the history of cities (and their renewal) who have left major legacies. This chapter is dedicated to Henry George (economic reformer and advocate for justice) (Plate 1).

Plate 1 Dedication Henry George. (Source: Wikipedia 2020)

Henry George *(1839–1897) was a popular economic reformer and advocate for justice in the late nineteenth century and visited many countries as far away as Australia. His book Progress & Poverty (1879) sold millions of copies worldwide, probably more than most books before that time. George spelt out how land ownership laws allow 'the* *rich to live off the poor'. Two testimonials to Henry George sum up his contribution to land value capturing (also called 'land value sharing'). The first comment is from Australia, where George visited during this time to see his ideas taken up by local governments in land rating policies:*

George's simple but emotionally inspired writings alerted people around the world (the Georgist movement) to fairer land development approaches. Proof of his popularity in Australia saw some 10,000 people attend George's inspirational speech at the Melbourne Exhibition Centre in 1890. Similar numbers followed his talks around the country. (Source: Prosper Australia web 2020)

The second testimonial explains a central aspect of Georgism as being examined today around the world:

George gained support by making economics understandable to the average person. The movement of Georgism is partly based on the principle that economic value derived from land should belong equally to all members of society. Georgism investigates the paradox of increasing inequality and poverty amid economic and technological progress. (Source: Wikipedia 2020)

Dedications Summary

There are many people who have left legacies contributing to the way cities can renew for the common benefit of all. Cities and regions are always evolving and attract leaders who see the potential in advancing their city or region. Each of the book's chapters is dedicated to two of these individuals. They are listed here, with a biographical note included in each of the respective chapters.

Front of Book: Henry George (economic reformer and advocate for justice)

Chapter 1: Sir John Bradfield (engineer) and Walter Burley Griffin (architect and landscape architect)

Chapter 2: Florence Mary Taylor (architect and engineer) and Neville Wran (NSW premier)

Chapter 3: Mum Shirl (social worker and community activist) and Colin James (architect and housing innovator)

Chapter 4: Sir Joseph Banks (naturalist and botanist) and Kevin Moss (parliamentarian NSW)

Chapter 5: Brian McGowan (politician) and Phyllis Bennett (teacher and arts advocate)

Chapter 6: Joyce Cummings (mayor and civic worker) and Greg Heys (mayor and sustainable cities advocate)

Chapter 7: Patrick Geddes (city planner and philanthropist) and Hebert Gans (sociologist and social planner)

Acknowledgements

I acknowledge firstly the contributions of residents, elected officers and government workers of Greater Sydney, City of Sydney, City of Canterbury and Bankstown, Central Coast Council (covering Gosford and Wyong Districts), and the City of Newcastle. Thanks also to the elected members and officers of NSW State and Federal governments for contributing information on the above centres.

At the University of Newcastle (Australia), thanks for consistent support from the academic and administrative staff. A thank you to the administrative unit of Newcastle School of Business (Gillian Armstrong and staff) and the school's senior lecturer Darren McKay. Thanks also to the School of Environmental and Life Sciences (Ourimbah Campus) academic staff Dr. Salim Momtaz and Simon and administrative staff Pam Steenkamp, Nicole Day and Fiona Adams.

Thanks to the town and environmental planners who contributed with comments on book structure, which includes: Bob Abnett, Kevin Armstrong, David Holland, and Rolf Fenner. Added thanks for comments to (a–z): Peter Adderley, Guy Caruana, John Desborough, Mark Ellis, Margaret O'Toole, Phil Stroud and John Wiggin.

I also wish to thank the following individuals and organisations: Joffre Balce (Association of Good Government, Sydney) and Mark Sullivan (Henry George Society, New York). On the book consultation side, I thank Habitat Association for Arts and Environment Inc. (HAAE) chairperson David Holland and public officer Kevin Armstrong along with members: (a–z): Ruth Dickson, Margaret O'Toole, and Diane Rauscher. Thanks also to book supporters: Susan Ferrandiz (historian, Portersville, Pennsylvania); Margo, Dylan and Marlon Levenspiel, Joyce Martin, and Richard Jr. Martin; Philip and Lenore Rauscher; Anna and Kevin Rochford (and Brooklyn); Paulie and Melissa Schmieder; and Maree and Michael Wheelahan (and Hugo and Portia). Special thanks to my wife Diane, for persevering with my questions and always encouraging.

Finally, I hope the book makes a contribution towards planning sustainable cities and regions, including: Greater Sydney; Central Coast; Greater Newcastle's cities of Newcastle, Lake Macquarie, Cessnock, Maitland and Shire of Port Stephens; the wider Australia; and beyond.

Contents

About the Author

Raymond Charles Rauscher Raymond Charles Rauscher is a conjoint lecturer at the University of Newcastle, Australia, and a director of Habitat Association for Arts and Environment Inc. Ray was born in Brooklyn, New York City (1943), and graduated from the City College of New York (Bachelor of Engineering (Civil), 1966). He later completed part one of a master's degree course at the University of Michigan (research on Detroit entitled *A Solution to the Urban Crisis: Proposal for the Creation of Region Serving New Cities* (unpublished), 1969). Ray completed his master's degree in town and country planning at Sydney University (1971), including a thesis *Community Response to a Redevelopment Proposal* (Sydney) (University of Sydney Library Microfilm Department 1971). The thesis covered planning conflicts in Erskineville (Sydney) and a framework to resolve urban planning conflicts. Delving into the subject of sustainable urban planning (SUP), Ray completed a PhD (2009) at the University of Newcastle, Australia.

Following further research, Ray published *Sustainable Communities: A Framework for Planning* (Springer 2014). Moving to address urban renewal, Ray completed research and published *Brooklyn's Bushwick – Urban Renewal in New York, USA* (Springer 2014). Researching further, Ray then examined urban planning (from the 1970s) in inner city neighbourhoods of Sydney, publishing *Sustainable Neighbourhoods in Australia – City of Sydney Urban Change* (Springer

2015). Moving from inner city planning, Ray then researched how cities were coping with today's impacts of global changes (i.e. population and environmental impacts) and published Cit*ies in Global Transition – Creating Sustainable Communities, Australia* (Springer 2016). The instance of these global impacts on neighbourhoods led Ray to commence research on the extent that cities, in planning neighbourhoods, address sustainable city principles. On further researching this question (with fieldwork over nine New York neighbourhoods in 2014), Ray completed *New York Neighborhoods – Addressing Sustainable City Principles*) (Springer 2017).

Ray commenced research into value capture after seeing a common need among cities and regions to define and act on the economics and design of urban renewal. To tackle this, he developed a model under the title Value Capture Planning (VCP). His research led him to examine cities in Greater Sydney and regions, including Sydney, Canterbury Bankstown, Gosford and Greater Newcastle. Ray would appreciate hearing from any reader on his/her views on this book's theme of sustainable cities and value capture planning (VCP) (email address – rayc.rauscher@gmail.com). Finally, any of the above books as noted can be accessed via a Google search 'ray rauscher books'.

Abbreviations

CBD	Central Business District
CERA	Central to Eveleigh Renewal Area
DCP	Development Control Plan
DPIE	Department of Planning, Industry and Environment (NSW)
EPA	Environment Planning Assessment Act
ESD	Ecologically Sustainable Development
FSR	Floor Space Ratio
GCCR	Gosford City Centre Revitalisation
GNMA	Greater Newcastle Metropolitan Area
GPO	General Post Office
GSC	Greater Sydney Commission
GSRP	Greater Sydney Region Plan
HCCRDC	Hunter Central Coast Regional Development Corporation
ICLEI	International Council for Local Environmental Initiatives
LA21	Local Agenda 21
LEP	Local Environment Plan
LVC	Land Value Capturing
LVS	Land Value Sharing
MTR	Mass Transit Rail (Hong Kong)
NCRA	Newcastle City Renewal Area
NCRS	Newcastle City Renewal Strategy
NGO	Non-governmental Organisation
NSW	New South Wales
NWRP	Newcastle West Renewal Precinct
PPP	Private Pubic Partnerships
SBURA	Sydenham to Bankstown Urban Renewal Area
SBURC	Sydenham to Bankstown Urban Renewal Corridor
SECS	Somersby to Erina Corridor Strategy
SEPP	State Environment Planning Policy
SIC	Special Infrastructure Contribution
SM	Sydney Metro

SMC	Sydney Metro City
SMS	Sydney Metro Southwest
SMN	Sydney Metro Northwest
SUP	Sustainable Urban Planning
UN	United Nations
UTA	Urban Taskforce Australia
VC	Value Capture
VCD	Value Capture Development
VCP	Value Capture Planning
VPA	Voluntary Planning Agreement
WHO	World Health Organization

List of Addendums

List of Plates

List of Tables

Part I
City and Region Renewal and Value Capture Planning Model

Part I of the book provides a background to governments adopting value capture and introduces the value capture planning (VCP) model. The focus is on cities and regions adopting value capture policies to achieve: equitable housing; public and open spaces; and, sustainable transport. **Chapter** 1 outlines planning for renewing cities and how value capture is applied. On opening the chapter, the implications of the post coronavirus era on the renewal of cities is reviewed. The chapter then shows how value capture works, providing a wide range of worldwide examples. The chapter also explains the roles of government bodies, developers and the community in value capture programs. A VCP model is developed in **Chap.** 2 along with the selection of four case study area cities.

Chapter 1
City and Region Renewal and Value Capture Planning

Abstract The chapter outlines planning for renewing cities and regions using value capture (VC). Before looking at VC more closely, the implications of the coronavirus pandemic (2020) for future urban planning (including value capture) are reviewed (i.e. housing, open spaces, and transport). The focus of the chapter then moves to cities adopting VC policies to guide renewal of parts of those cities. Those governments that use VC within their planning are then examined. It's explained how VC policies are triggered by government decisions. These include for example: land up-zoning and designation of major government infrastructure projects. The chapter then addresses key aspects of VC planning, including: (1) background to value capture; (2) applying value capture; (3) innovations in value capture; (4) value capture approaches; (5) value capture in different countries; (6) contrasting views on value capture; and, (7) variations and confusions about value capture. Finally, the chapter explains how value capture is administered through value capture planning (VCP) programs, including engaging the governing body, developers and the community.

Dedication There are many people in the history of cities and renewal who have left major legacies. This chapter is dedicated to two of them (Plate 1.1), Sir John Bradfield (engineer) and Walter Burley Griffin (architect and landscape architect).

John Job Crew Bradfield (1867–1943) was an Australian engineer best known as the chief proponent of the Sydney Harbour Bridge, of which he oversaw both the design and construction. He was the first recipient of an engineering doctorate from the University of Sydney. Other notable projects with which he was associated include the Cataract Dam (completed 1907), the Burrinjuck Dam (completed 1928), and Brisbane's Story Bridge (completed 1940). The Harbour Bridge formed only one component of the City Circle, Bradfield's grand scheme for the railways of central Sydney, a modified version of which was completed after his death (Source Wikipedia 2020).

Walter Burley Griffin (1876–1937) was an American architect and landscape architect. He is noted for designing Canberra, Australia's capital city. He has been credited with the development of the L-shaped floor plan, the carport and an innovative use of reinforced concrete. Influenced by the Chicago-based Prairie School, Griffin developed a unique modern style. He worked in partnership with his wife Marion Mahoney Griffin. In 28 years they designed over 350 buildings, landscape and urban-design projects as well as designing construction materials, interiors, furniture and other household items (Source Wikipedia 2020).

R. Rauscher, *Renewing Cities with Value Capture Planning*,
https://doi.org/10.1007/978-3-030-62958-8_1

Plate 1.1 Dedications: Sir John Bradfield and Walter Burley Griffin

1.1 City and Region Renewal in Post Coronavirus Era

(2020) many cities around the world are looking for ways to plan urban renewal using land value capture (herein called 'value capture' or VC). The focus of this planning is on land that is likely to result in increased value for owners (including developers) as a result of government planning decisions. Cities utilising VC will usually also seek to integrate VC into sustainable urban planning (SUP) principles being applied in development (Rauscher and Momtaz 2014, 2015; Rauscher 2018). One means of achieving these principles is adopting VC programs. This is done partly through developer provision and/or VC levies associated with development. The outcome of developer provisions and levies enable governments to meet public and infrastructure needs associated with developments. Before looking at VC more closely, the implications for future urban planning (including VC) of the coronavirus pandemic (2020) is reviewed.

The coronavirus (COVID 19) swept upon the world and soon ravished sections of cities and regions at hot spots. It appears that it was the denser urban sections of cities that felt the brunt of the virus. This was especially the case in poorer sections of cities, where housing overcrowding may have contributed to the virus spread. In terms of this book's theme (renewing cities with value capture), the COVID 19 virus's impact on urban centers raises questions. Some of these questions evolve around the resilience of the city structure in catering to the needs of its population when disasters such as infectious diseases strike. In a virus strike the community responses would include: a resilience to contain the virus spread; and, suppression and/or eradication within the multitude of city communities. A general background on infectious disease and pandemics relating to cities is a starting point to explore the subject. What follows is a brief summary (researchers noted) of the factors to consider in minimizing impacts of future infectious diseases within urban settings. This summary is laid out in four parts: (1) Pandemics and Cities; (2) Health Factors in Designing and Renewing Cities; (3) Infectious Diseases, Global Warming and Urbanization; and, (4) Housing Design in Post Virus Era.

1.1.1 Pandemics and Cities

Civilisation has experienced many pandemics. Coronavirus joins the list as one that has wreaked havoc across the globe. There is a history of individuals and groups in the past tackling the impact of pandemics on urban communities and how city structures cope. A definition of a pandemic, to start with, is as follows:

> A pandemic is a disease epidemic that has spread across a large region, for instance multiple continents, or worldwide. Throughout history there have been a number of disease pandemics such as smallpox and tuberculosis. One of the most devastating pandemics was the Black Death, which killed an estimated 75–200 million people in the 14th century. Current pandemics include HIV/AIDS and the 2019 coronavirus disease. Other notable pandemics include the 1918 influenza pandemic (Spanish flu) and the 2009 flu pandemic (H1N1) (Source Wikipedia 2020).

Knowing the background of the coronavirus is important in future renewal planning of cities. Karlen (1995) describes why this virus was so lethal:

> Coronaviruses (CoV) are a large family of viruses that cause illness ranging from the common cold to more severe diseases such as Middle East Respiratory Syndrome (MERS-CoV) and Severe Acute Respiratory Syndrome (SARS-CoV).

It was soon learned in early 2020 that a new strain of coronavirus (SARS-CoV-2) caused coronavirus disease (or COVID-19). This disease was declared a pandemic by the World Health Organisation (WHO) on 11 March 2020. Previously there had been a number of studies on pandemics affecting city and regional populations. Being prescient, The Club of Rome (2019), launched *The Planetary Emergency Plan* on the sidelines of the UN Climate Action Summit in New York. The Club of Rome comments on health and environment noting:

> The Plan provides a set of key policy levers addressing the cross-cutting challenges of climate change, biodiversity loss and human health and well-being. It outlines a vision of transformation and regeneration. It is a roadmap for governments to set in motion a decade in which the development path of our planet is steered onto a new direction.

The Cub of Rome states that the Plan was based on a ten-year action framework and outlined calls for transformative action from 2020 onwards. The Plan also calls for the longer-term 'systemic shifts needed to truly emerge from the planetary emergency'.

1.1.2 *Health Factors in Designing and Renewing Cities*

Following the above background on pandemics and their impact on cities, the positioning of health factors in designing and renewing cities needs examination. Specifically, relationships between infectious disease spread and cities include variables of: (1) urban densities; (2) poorer sections of cities; (3) and, housing design. What follows is a brief summary (research based) on these relationships.

Ian Klaus (2020) comments on minimizing the risk of infectious diseases:

> People have redesigned cities, infrastructure, architecture, and interiors all in the name of minimizing the risk of infectious disease. We are constantly making adaptations to our cities, neighborhoods, and homes. A portion of that adaptation is designing within a medical framework. The question is raised how we can use the built environment as a way to reduce epidemic spreads? Thus can we respond to infectious disease by redesigning physical spaces?

Klaus continues in commenting on planning and advancing public health:

> The burgeoning field of urban planning used public health to advance its ideas in the 19th century. Many 20th century modernist architects viewed their buildings as a type of medicine model. For example, architect Le Corbusier viewed light and air as being medicinal. In time, public health was used to rationalize slum clearance programs across many countries (especially after WW2 and into the 1950s).

Tracy Hadden Loh (ET all) (Brookings Institute 2020) comments on the relationship between the coronavirus' origin (and its diffusion) and rapid global urban growth. She argues that this urban growth facilitates virus transmission from animals to humans, person to person, and country to country. She then addresses the physical connectivity of people and authorities' ability to cope with disease transmission:

> Given this virus transmission the connectivity of people and places may matter now more than ever. Local responses to the pandemic are revealing that in the midst of mandated distancing, the economic, physical, social, and civic structures of communities significantly influenced our ability to cope with the immediate crisis.
>
> Ref. web – https://www.brookings.edu/blog/the-avenue/2020/03/25/the-qualities-that-imperil-urban-places-during-covid-19-are-also-the-keys-to-recovery/

How do we thus build healthy (and still vibrant) places? Robert M. Bass (2020) (Center for Transformative Place-making) released a framework (in looking at viruses) describing the interconnected elements needed in order to build healthy and vibrant places. These places were being tested in coronavirus 2020 real time. Bass comments on his framework:

> Despite serious downsides, the physical qualities of densely connected places can also make life amid social distancing more liveable. Walkable and bikeable places are easing the isolation of quarantines and enabling people to move about. Public parks are giving people a respite for fresh air and exercise while maintaining social distance. With fragmented federal and state-level public health responses to the crisis, local governance structures are more essential than ever. The actions of civic groups, from neighborhood associations to business improvement districts, are reinforcing. The locally organized, inclusive, and networked communities will bring benefits, resources, and an organized voice to residents. The

> Center for Transformative Place-making concludes that everyone has the right to live in a great place, and the right to make the place better.

Looking towards prevention, however, how much disaster planning is required within any city or community? Michael Berkowitz (2020) examines disaster planning and how cities weather them. As the former executive director of the non-profit consultancy 100 Resilient Cities (and founding principal at Resilient Cities Catalyst) Berkowitz has worked around the world in planning for impacts of hurricanes, droughts, earthquakes, terrorist attacks, mass shootings, disease outbreaks, and other social shocks. Berkowitz connects disasters with community urban resilience, and herein outlines what a coronavirus recovery could look like, noting:

> Urban resilience is the ability or capacity of a city to survive and thrive in the face of disaster, any kind of disaster. It turns out that the capacities that cities, states or nations, need to survive and thrive in the face of threats are pretty broad. It includes good infrastructure that promotes mobility and sustainable transportation. It's also about cohesive communities where neighbors check in on neighbors. It's a diverse economy with a strong middle-class jobs base. It's good governance with multiple stakeholders at a decision-making table. All of those things help communities overcome whatever the next crisis might be. We should also set new goals. We have the tools to think about economic development, public health, biodiversity, and flood control all at once. We have the beginnings of examples of what those types of projects look like, such as building large swaths of green infrastructure to protect cities from sea level rise, like the BIG U in Manhattan.

Looking further into the question of pandemics related to living patterns (and future city design), Planetizen (2020) notes:

> Confined to the relatively small spaces of homes and apartments, surrounded by immense cities and regions, questions are raised. These include whether living arrangements and development patterns are resilient and safe enough take on new meaning during a pandemic, and new anxieties. As news continues to break and the situation continues to change, the media will continue to debate the lessons of the pandemic, including ideas about urbanism. Some of the leading voices in urbanism, planning, and design media are already attempting to reconfigure a vision for the future of cities, now that the coronavirus has revealed so much about the ways we live.

Planetizen continues in addressing the future role of planners:

> What can we expect when the pandemic is over? This question serves as a context for challenges facing communities as the worst effects of the pandemic start to show in hospitals and unemployment numbers. The connections between the traditional role of planning and the future of planning needs attention (i.e. more attentive to accommodating public health). Planners will be essential in the hard work of answering questions about the public realm, mobility, social isolation, and local and regional leadership.
>
> Ref. web – https://www.planetizen.com/blogs/108814-debating-future-cities-and-urban-density-after-pandemic

Hence, can we spell out what is the role of planners and architects in designing healthy cities and meeting impacts of future pandemics? Ann Forsyth (2020) reflects on COVID-19's impact on the future of urban life as follows:

> In the context of COVID-19, some have questioned the future of urban life. COVID-19 is emptying out public transportation in many places, but transportation is already in a transition period due to automation. The key health crisis from COVID-19 is likely to appear in

> more crowed settlements without adequate water supplies and sanitation (in both urban and rural settings). These have been a focus of public health concern for a long time. While cities will not be eliminated, a long period of suppression may well change patterns of urban life.

Forsyth then looks at moving from a disaster to addressing a wider range of health issues in future urban planning, noting:

> For the past decades, those looking at the intersections of planning, design, and public health have focused less on infectious diseases and more on three other areas: chronic disease, hazards, and the vulnerable. For *chronic diseases,* those lasting a year or more, the environment can provide options for healthy behaviors such as physical activity or mental restoration. For *hazards,* such as climate change, planners and designers need to address flooding, droughts, and climate-led migration. And for *the vulnerable*, the environment needs to focus on those who are old, young, have pre-existing conditions, or have low incomes. The current pandemic brings the question of designing for infectious diseases back to the forefront, however, and raises important questions for future research and practice.
>
> Ref. web – https://www.gsd.harvard.edu/2020/03/what-role-do-planning-and-design-play-in-a-pandemic-ann-forsyth-reflects-on-covid-19s-impact-on-the-future-of-urban-life/

Kinder Rice (2020) also looks at the long-term effects of the Covid-19 crisis (taking the US as an example, however her comments could relate to any urban area). She writes about urbanism as a driver of social and economic life (in spite of pandemics):

> Urbanism is one of the most important drivers of both social and economic life in the United States. Even if most people live in the suburbs, economic innovations prospering the most these days are in the city. The most critical component is proximity. In the world of technology, much has been made of the way people bump into each other almost randomly in dense job centers and exchange ideas. When we talk about a revival of "urbanism," what we are really saying is people have rediscovered the value of proximity. Of course, now we are learning (given viruses) that proximity isn't always a good thing.

Rice then addresses how cities are resilient and can accommodate lessons from the Covid-19 pandemic (2020):

> The fact that the virus has ripped through New York City (America's densest city) has brought the urban naysayers out of the woodwork, claiming cities are the root of pretty much every problem. Yet cities are much cleaner and safer than they were a century ago. That's because cities are by their nature adaptable organisms. More efficient than rural areas and more flexible than suburb. Cities are constantly reinventing themselves. The world after COVID-19 will be different, as it is after any disaster. And COVID-19 will accelerate changes that have been brewing in cities for a long time. The result will be a new kind of city, different than what we have seen before. This will be a city that should be able to withstand shocks like COVID-19 in a sturdier fashion.
>
> Ref. web – https://kinder.rice.edu/urbanedge/2020/03/26/what-our-cities-will-look-after-coronavirus-pandemic

Finally, Tahj Rosmarin (2020) focuses on the importance of public spaces and communal values in future planning of cities (post pandemic), commenting:

> In modern cities, our public spaces represent our shared values. They are our common assets, owned, maintained and used by all members of our society. The outbreak of corona-

> virus and its immediate impacts, such as social distancing, have raised many questions about the role of public space in such times. Our trust in public spaces will need to be rebuilt if we are to rebuild our economy and our society. In the face of generational issues such as climate change and homelessness, there is hope coronavirus might actually offer us an opportunity to radically reassess our communal values.

Rosmarin moves on to address the role of public spaces in accommodating public health parameters:

> The squalid conditions of the late 19th-century city were swept away by large-scale public infrastructure investments in health and safety such as centralized water, sewer and lighting systems. Moving on, during the suburban era, we tried to solve most of our land-use planning problems by putting more space between people and buildings. But in cities, that's not possible. Instead of spacing our way out of problems, we have to design our way out. The threat of infectious disease is likely to ramp up urban design as a solution. This could include, for example, separation of people in public spaces like restaurants and parks.

Rosmarin concludes on addressing public ownership of spaces:

> So, it's reasonable to assume we'll begin to see small changes in urban design that separate people a little more and help make it easier to protect them. Maybe the best analogy is the "defensible space" movement started by urban planner Oscar Newman's view was that all space belonged to somebody. By giving residents a sense of ownership of these spaces, he contended, gangs wouldn't take them over. It was a revolutionary theory in urban design that changed everything. This showed that instead of fleeing cities it meant standing firm and re-designing for a better solution.

Having addressed health factors in designing cites, the subject of infectious diseases connection to global warming and urbanisation is examined next.

1.1.3 *Infectious Diseases, Global Warming and Urbanization*

The Covid-19 pandemic (2020) has raised questions on the connections (in designing and renewing cities) between infectious diseases, global warming and urbanization. On a city planning scale Sue Surkes (2020) addresses what is the likely effect of climate changes on incubating infectious diseases (such as the coronavirus 2020). Surkes firstly notes that population growth will increase likely infectious diseases, given the loss of natural habitats. With this loss (she argues) wild animals come more into contact with humans, thus easing the way for infectious diseases.

> Ref. web – https://www.timesofisrael.com/viruses-expected-to-increase-with-global-warming-expert/

Surkes argues that climate change impacts on human health poses risks and will require appropriate national and international responses. The World Health Organisation (WHO) (2020) notes that worldwide there is an apparent increase in many infectious diseases. This includes some newly-circulating ones (HIV/AIDS, Hantavirus, Hepatitis C, SARS, etc.). This reflects (WHO notes) the combined impacts of rapid changes in: demographic, environmental, social, and technological ways-of-living.

The WHO explains the two categories of research into the linkages between climatic conditions and infectious disease transmission. These are: (1). evidence from the recent past of associations between climate variability and infectious disease occurrence; (2). early indicators of already emerging infectious disease impacts of long-term climate change. The WHO goes on (given the above) to suggest governments need to create predictive models to estimate the future burden of infectious disease under projected climate change scenarios. The WHO concludes that changes in infectious disease transmission patterns are a likely major consequence of climate change. The WHO admits that the world community needs to learn more about the underlying complex causal relationships, and apply this information to the prediction of future impacts. This means using more complete, better validated, and integrated models (says the WHO).

Ref. web – https://www.who.int/globalchange/summary/en/index5.html

Moving to nation and international challenges of climate change and infectious diseases, Climatewire (introducing Chelsea Harvey) (2020) raises the question of what could earth warming mean for pathogens like coronavirus? Harvey believes that generally, scientists expect to see changes in the timing, location and severity of disease outbreaks as global temperatures rise. She notes:

> As the earth warms we can expect to see changes in the timing, geography and intensity of disease outbreaks around the world. And some experts believe climate change, along with other environmental disturbances, could help facilitate the rise of more new diseases, like COVID-19. A paper in 2013 found that unusually warm winters tend to be followed by earlier, more severe flu seasons the next year. The researchers suggest this is because fewer people come down with the flu during warmer winters, leaving their immune systems more vulnerable the following year. The world battled in 2020 a pandemic that seemed potentially more dangerous than any other infectious disease in a century.

In conclusion, most of the above research on climate change reinforces a view that viruses can be expected to increase with global warming (urbanization spread and deforestation contributing).

1.1.4 Housing Design in Post Virus Era

Moving to a post virus era and future housing design, the question of urban collectivity and housing (i.e. vertical living) is raised. Eran Chen (2020) addresses this question, as follows:

> With over half the world's population inhabiting cities or densely populated areas, billions of people are currently residing in small spaces, disconnected from one another by brick, concrete and steel. Long before this global pandemic vertical living was seen to contribute to a state of isolation. Vertical living has turned us into observers, instead of participants. It is our design intention to use architecture to bridge communities and create spaces for connection that were lost in the decades of urban planning. The rush to accommodate took priority over the need for community.

Chen notes that architecture has the power to shape behaviour. Extreme conditions often clarify what otherwise is vague or uncertain. The nature of emergencies (Chen writes) is that they surface acute questions about our way of life, social structure and interactions. In addressing this question he states that density doesn't have to be the enemy. He notes that a well-designed community can be the solution. European villages, with their balconies and inner-courtyards and shared clotheslines, have a built-in layer of intimacy. Chen states that architecture has the power to revitalize communities in the same way, even if in a more modern form. He states that even a mammoth 2-block mini-city, comprised of 1 million square feet, and housing 4000 residents under a single roof, can feel like a neighborhood, light and airy and filled with song. Chen then addresses the concerns of today's loneliness and depression within our urban communities, as follows:

> What must we weigh from this shared pandemic? One of the biggest concerns today is one of acute loneliness and depression. When thinking about resiliency, we should be thinking about social resiliency as well. As an architect living in NYC, I have always put a great value on green space, indoor and outdoor space, and a connection to nature. We must address light and air quality and the ability to step outside and invite in a new perspective filled with smells and sounds. But it takes more than just a balcony to share such human experiences. People need to feel secure in their homes, have common "territory" that allows neighbors to see and hear one another. While amenities are often seen as a craze or a fad, we see them as the new social fabric. We see them as the new neighborhood. Green walls and inner courtyards aren't window dressing, they are a direct connection from your brain to the diurnal, indigenous instincts we have as humans on earth.

In addressing this home design question in the post coronavirus era, Chen continues:

> While the situation caused by the coronavirus is still unfolding, certain things are becoming self-evident. There must be a better way to arrange our homes in our increasingly dense cities where we can enjoy our privacy while acknowledging our neighbors. It is where we can all access outdoor spaces and feel the sunshine on our faces. We need amenities that aren't buried, but are treated as a luxury, that gives us the option of social engagement or time apart, that create nooks as well as recreation. This is what smart design means in the future. Design that meets all our needs, our human rights and instincts that speaks to the collective whole, and therefore the collective good.

There are many housing design aspects that city administrators and planners will be examining, post coronavirus 2020. Looking at the increasing high rise living, several studies (for example Mao 2015) suggest that during the outbreak of highly infectious disease, high-rise dwellers are at higher risk than people living in single or detached homes. Mao notes that those living in large apartment buildings worried about increased exposure to coronavirus. This was caused by concerns of the number of residents touching door handles and elevator buttons.

There are the mental health issues too (triggered by pandemics) centred on our style of habitation. Australian architect Kerry Clare (2016) states that high-rise living can isolate people from street life. According to Clare, people living in high-rise buildings have fewer chance encounters of street life. In many respects, Clare's position resonates with the thinking of New York's Jane Jacobs (1961). In *The Death and Life of Great American Cities*, Jacobs criticized aspects of high-rise living, especially when used as a low-income housing solution.

Clay Lucas (2016) Sydney-based architect researcher and Professor Kerry Clare (2016) have commented on some of the drawbacks (i.e. social isolation and diminished public realm amenity) of high rise living. Professor Clare questions why Australia is relying on high-rise to house its growing population, noting:

> Living in a high-rise building radically reduces the sort of chance encounters that lower-rise dwellings afford, as residents were on the street more often. High-rises also diminish people's participation in public spaces. High-rise towers are largely built during economic bubbles, and many are left empty as investor-owned properties.

Commenting, in addition, on the need for more outdoor spaces around high-rises Professor Buxton (2016) states:

> Melbourne (one of the top 10 locations worldwide for construction of high-rise towers).was building too many central-city high-rises. At the same time Melbourne's suburbs continue to sprawl outwards. Not enough medium density was being built in the middle-ring suburbs. Too many Australian apartments suited only singles or couples with no children. Many of these apartments were being lived in by short-term visitors, students or those staying only months in one spot. In looking at outdoor spaces around many high-rise buildings generally there is a need to have green infrastructure (i.e. large parks, natural landscapes, quality public streets and urban spaces).

Looking towards future high-rise designs, there are however other studies that suggest living higher rather than lower may, in fact, increase mortality rates. Panczak (2013) published in the *European Journal of Epidemiology* the mortality research results on people (1.5 million survey) living in buildings of four or more floors. Panczak found that to make high-rise living healthier more design innovations are needed. These could range from green roofs to the 'breathable' buildings associated with the Passive House movement.

1.1.4.1 Conclusion

Given this above summary of some of the factors to consider in designing and renewing cities to minimize impacts of future infectious diseases, governments will need to take notice. The areas for governments to examine (as reviewed) include: planning for healthy cities; reducing impacts of urbanization on global warming; and, ensuring future housing design is resilient (especially during pandemics). To accomplish this (and more) governments, development industry and the community will need to investigate the design and the financing of renewing cities. Hence, value capture planning (book subject) may offer a key avenue, as introduced next.

1.2 Value Capture Planning (VCP)

Having addressed above future directions of urban planning (in the post coronavirus era), the subject of value capture planning (VCP) is introduced. This planning tool could become more central to renewing cities, the subject of coming chapters. This introduction includes: (1). background to value capture; (2). applying value capture; (3). innovations in value capture; (4). value capture approaches; (5). value capture in different countries; (6). contrasting views on value capture; and, (7). variations and confusions about value capture. Each of these aspects of VCP will now be addressed, with multiple examples of projects utilizing VCP.

1.2.1 Background to Value Capture

There are a number of sources that explain aspects of value capture. Wikipedia summarises these aspects as follows:

> Urban planners and finance officials are interested in value capture mechanisms, given: 1. value capture offers a targeted method to finance infrastructure benefiting specific land; and, 2. these investments can generate private investment in the area, which will more widely benefit the city.
>
> Thus, for example, when a new subway station or highway interchange is installed, land near the new facility becomes more valuable. Investment in capital improvements to land can thereby generate capital investment in other nearby locations. This further increases land values. (*Source* Wikipedia General Summary 2020).

A number of studies have addressed value capture planning. Australia, given its current rapid urban renewal and city expansion plans, is a place where VC could be more extensively applied. Consult Australia (2013), in explaining value capture (VC), produced the report *Capturing Value*. The report focuses on the need for a wider program for governments' better meeting community and infrastructure needs generated by developments. The report firstly points out that many planning agencies around the world have adopted VC procedures (or other forms of funding public needs). Consult Australia notes that there is a complexity of funding sources for infrastructure projects in Australia. Infrastructure needs for a state such as New South Wales (NSW) are typically funded from a combination of Commonwealth (federal) grants, state taxes, council rates, user charges and development levies (s94A as examined below). This combination (report notes) leads to complications for any intergovernmental cost-benefit analysis and policy development in VC.

A table (Addendum 1.1) has been completed (by the author) to illustrate the breath and variety of value capture project applications around the world. Looking at an overseas project (#3 in Addendum), for example, the transport agencies and local businesses in England are applying VC to the Crossrail project (new rail line). Here VC funds collected over 30 years will be used to finance a quarter of the $29.6 billion project (that is $7.4b value shared). Within the table are varied VC projects listed by location and project type. Most projects include allied development of residential and commercial. Project types within the list (numbers of projects in brackets) include: heavy rail (and related developments) (5); land use rezonings (4); light rail (2); sports infrastructure (2); and, commercial and residential (1), The table shows the wide range of value capture projects. The most popular example cited is rail (heavy and light). The next most popular cited projects were those under land use rezonings (i.e. up-zonings). Further comments of all the projects listed will follow within the chapter.

Within Australia, VC can be distinguished from the more traditional forms of development contributions under the states. Within NSW for example the State administers (from 1970s) development contributions. This program is under Section 7.11 Contributions (was noted as s94) and Section 7.12 Levies (was noted as s94A) of the Environmental Planning and Assessment (EPA) Act (NSW 1979) https://www.planning.nsw.gov.au/local-infrastructure-contributions-policy

In NSW development applications are thus often approved subject to conditions imposed under these two sections of the Act. This requires development contributions to be made towards the cost of providing site related infrastructure and facilities. This contribution is based on a nexus principle of connecting a development with the triggering of related needs. Thus a development contribution or levy is for an individual site development proposal, and thus not based on the principles of value capture. The NSW Government summarizes development contributions and levies within the Act as follows:

1. Contributions for local infrastructure, also known as developer contributions, are charged by councils when new development occurs. They are used to provide infrastructure to support development, including open space, parks, community facilities, local roads, footpaths, storm water drainage and traffic management.
2. Section 7.11 contributions are charged where there is a demonstrated link between the development and the infrastructure to be funded. Councils prepare contributions plans which specify what infrastructure will be provided and approximately how much it will cost. This is used to calculate a contribution rate, usually charged per dwelling or per square metre. Councils that want to charge a contributions rate above the threshold set by the Minister must submit their plans to IPART for independent review.
3. Section 7.12 levies is an alternative to s7.11 contributions, charged as a percentage of the estimated cost of the development. The maximum percentage that can be charged in most areas is 1% of the cost of develoment, although there are a small number of areas that charge a higher percentage.
4. The local infrastructure contributions system is administered by local government as they are best placed to understand the needs of their communities. Questions about specific contributions plans and contribution rates should be addressed to the relevant council.
5. The Department sets the policy framework under which councils collect and administer the contributions. This includes legislation, directions made by the Minister for Planning and Public Spaces and practice notes.
6. The Department notes that Section 7.11 EPA Act (1979) contributions are subject to the 'nexus requirement' (a public need stemming from development) (Planning and Assessment Regulation 2000) (DPIE 2000).

The NSW Department of Planning, Industry and Environment (DPIE) (2016) issued a draft Practice Note and draft Planning Circular on the use of planning agreements to fund items of infrastructure. The draft Practice Note encourages councils to consider planning agreements supporting a "reasonable share" of development profit. This is an incentive within contribution requirements. It is also an encouragement for contributions in the form of works-in kind in response to an infrastructure plan prepared by a Council. The Circular also addresses the use of voluntary planning agreements (VPAs) for funding infrastructure and other public benefits. This extends beyond the scope of contributions (Section 7.11) and levies (Section 7.12) of the EPA Act (1979). The Department outlines key provisions of VPAs, including:

1. VPAs facilitate development contributions for a variety of public purposes
2. Infrastructure and public benefit is likely to be delivered in a more comprehensive way if undertaken in tandem with broad strategic planning, rather than determined at the site level.
3. VPAs are suited to large development sites where there are "clear benefits in the managed delivery of public benefits in association with development"
4. The process and procedure for negotiation of VPAs should be identified by a planning authority and made clear to a developer and VPAs should provide a public benefit that has a clear link to the development.
5. Consideration of a planning proposal should not be premised on the financial outcome of a VPA.
6. Planning authorities should identify when money received under various planning agreements is to be pooled and applied towards the provision of public benefit.

The Circular then addresses planning agreements more broadly, noting:

1. Planning agreements provide a flexible framework under which state and local councils can share in the provision of infrastructure in new release areas or in major urban redevelopment projects.
2. Planning agreements do not need to meet a 'nexus requirement' (noted above) because they are by nature voluntary (unlike Section 7.11 contributions which are compulsory).
3. A planning agreement can wholly or partly exclude the application of local infrastructure contributions (by councils) or special infrastructure contributions (by state government). This gives the planning authority scope for negotiating trade-offs in an agreement. This means financial, social and environmental costs and benefits of the development can be redistributed through an agreement.
4. While planning agreements should complement other contributions mechanisms, they should not be used as defacto substitutes for contributions plans.
5. The concept of value capture is supported, however the provision of planning benefits for the community should be "capturing part of development's profit". The percentage capture should be a "reasonable share of development profit", ensuring the developer's share of profit does not fall below that which is required for the development to be feasible.
6. Planning agreements are not to be used as a mechanism to capture "windfall gain".

7. Planning agreements are evidence based, preferably peer reviewed and used as a mechanism to introduce agreed public benefit, developed through the processes of strategic planning and community consultation.
8. A proposed development needs to provide opportunity for public benefit and infrastructure (including affordable housing) to be delivered by development with regard to the fair apportionment of costs.
9. The method of apportioning infrastructure costs needs to be clear, justified and ensure the developer has an entitlement to profit that enables development to proceed.
10. Proper investigation and consideration of development feasibility and capacity to pay needs to be carried out. This should preferably be on an "open-book basis".
11. If a planning agreement is to be linked to planning incentives, density bonuses, planning trade-offs, etc., details of the relevant scheme and its implementation should preferably be contained in an environmental planning instrument or development control plan.
12. In the case of a bonus scheme, planning authorities should carry out public consultation, consider the apportionment of funding, examine the feasibility impact and determine the need for the infrastructure.
13. Planning proposals, development consents or modifications should be considered on their merits, with the unwillingness of a developer to enter into a planning agreement related to land value increase not a reason to refuse a proposal.

Further, in a report by AEC (2016) to Newcastle City Council addresses application of value capture principles within Council's (2018) Wickham Master Plan (Chap. 6).

https://www.newcastle.nsw.gov.au/getattachment/Development/Land-Use-Planning/Wickham-Master-Plan-Project/Wickham-Masterplan-Economic-Analysis-Final.pdf.aspx?lang=en-AU

A number of value capture mechanisms were outlined in the report and would be applicable to many planning authorities as a guide when considering adopting value capture programs, including:

1. One way of attracting developer interest to an area is to increase the height and floor space ratio (FSR) controls. This, however, often increases landowner expectations and can result in a scale of development that is not financially viable.
2. The financial benefit resulting from a development of bonus floor space is enjoyed both by the landowner and developer. The landowner benefits from an increase in land value and the developer benefits from an increase in development profit. In many cases the landowner and developer are the same party.
3. By levying a contribution rate for bonus floor space that captures a proportion of the land value uplift, both developer and landowner benefit. Often only a proportion of value uplift is captured (say 50%). The remaining value created (50%) benefits the landowner while the developer benefits from a higher profit for a larger development.
4. This mechanism (value capture) is only viable where the prices paid for development sites reflect the contribution required for bonus floor space, i.e. that a developer does not overpay for a site.

5. The City of Sydney and City of Ryde (in respect to value capture) codify and detail bonus floor space contributions so that the relevant parties are informed at the outset and able to make informed decisions at the time of site purchase (i.e. not to overpay for a site).
6. A rezoning or up zoning will usually lead to increased landowner expectations (which then influences the price paid by a developer to assemble a development site).
7. Owing to statutory limitations not all infrastructure can be funded by Section Sections 7.11 contributions or Section 7.12 levies.
8. The use and role of incentive based mechanisms for infrastructure funding are important.
9. There are only a few incentive-based infrastructure funding mechanisms that are codified in NSW. Those few include Green Square Community Infrastructure contributions (formerly known as the Bonus FSR Contributions System) and Ryde LEP 2014 (Amendment 1) Macquarie Corridor (formerly Macquarie Park Bonus FSR Contributions Scheme). These generally seek to leverage:
    (a) Density bonuses and/or planning concessions in a council (i.e. LEP) or State (i.e. State Environmental Planning Policy) (SEPP)
    (b) Capture of planning gain/value uplift associated with a rezoning or increased density, wherein contributions to public benefit are sought.
    (c) Widespread market/industry acceptances evidenced by high levels of take-up in Green Square.
    (d) Bonus contributions for residential markets while recognising that retail/commercial markets are more fragile and have less tolerance to additional contributions.

Finally, looking at value capture and provisions of affordable housing, the City of Sydney (2012) operates the Green Square Community Infrastructure incentive provisions.

https://www.cityofsydney.nsw.gov.au/__data/assets/pdf_file/0004/128074/GreenSquareCommunityFloorspaceGuidelinesDec2012.pdf

These provisions led to new planning controls allowing for the transition of employment lands in Green Square from traditional industrial uses to a diverse range of other uses, including residential. The Employment Lands Affordable Housing Program came out of these planning controls and encourages the provision of affordable rental housing. The program includes: application of a levy to fund new affordable rental housing; and, makes allowances for residential sites to deliver affordable rental housing.

Looking to the US and beyond Australia, Sections 7.11 and 7.12 (EPA 1979) mirrors impact fees and exactions mechanisms in parts of that country. Likewise, the Community Infrastructure Levy (CIL) operating in the United Kingdom has a similar basis as Section 7.12. The UK levy thus (as with Section 7.12) provides a pool of funds to the planning authority to meet identified needs created by development projects (however not based on value capture).

1.2.2 Applying Value Capture

What are the challenges of applying value capture (VC), as examined next? To clarify this further, Consult Australia (noted above) has put together *a Land Value Capturing Roadmap* Consult Australia (2016) (herein called the Roadmap (see link below).

http://www.consultaustralia.com.au/docs/default-source/cities-urban-development/value-capture-roadmap/value-capture-roadmap-as-web.pdf?sfvrsn=2

The Roadmap highlights opportunities and challenges of VC funding methods. This document utilises NSW as a case study to demonstrate the case for VC. The Roadmap notes that evidence from around the world demonstrates that well-conceived infrastructure investment creates new economic activity. The adoption and implementation of VC is outlined in Addendum 1.2. This includes: (1). explaining the State context; (2). pilot programs; (3). projects; and, (4). value capture legislation.

Consult Australia (2015) notes that, looking further at Australia, land value increases (and VC application) are related, for example, to improvements to Sydney's suburban rail network. Land values nearly tripled at Epping for single dwelling properties near the new expanded rail station (increases from an average of $1.2 million to over $3 million over several years in the early 2000s). Consult Australia (2015) thus points out there are private financial windfalls as a direct result of the public's investment in transport infrastructure. Consult Australia notes that there is, however, no equitable mechanism existing in any Australian state or territory to capture this value. This would then flow the value to authorities responsible for public provisions). Consult Australia further explains that cities and states are not obtaining a financial benefit from their infrastructure investments. It's argued that the public is in effect paying an inflated price for land around transport infrastructure as a result of this public investment. Consult Australia notes this uplift in value at times solely benefits nearby property owners and investors (not the public). Inflated land costs are often then paid by the State, and then passed on by the State in the form of higher taxes on taxpayers. As a result consumers then also pay higher public transport fares and higher housing costs. There is a clear case here to be looked at from the taxpayer's position. Innovations in value capture could be also be looked to assist governments, developers and taxpayers, examined next.

1.2.3 Innovations in Value Capture

Examining value capture in Australia further, Consult Australia's Joe Langley (2015) comments on innovations. He writes that cash-strapped governments have realised the need to find more innovative ways, for example, to pay for infrastructure such as public transport. Mr. Langley calculates that $6 billion could be saved if value capture methods were used (2020) on the $60 billion of projects which the

NSW state government is funding. Currently government funds come from a long term lease to a private company of the State's electrical system of poles and wires.

Langley notes the Mass Transit Rail's (MTR) run between Rouse Hill and Chatswood (commenced August 2019). The line will be extended to Bankstown through the CBD and Waterloo (Chap. 3) and Sydenham (Chap. 4) by 2024. It is understood (for example) that the contracts generally miss an opportunity for comprehensive property development rights returning to the public (addressed further in the chapter (under Value Capture Approaches).

More recently, the Sydney inner suburb of Waterloo (noted above as part of the extension of Metro to Bankstown) was nominated for a new underground Metro station (construction commenced in 2019). Above the station will see four new high rise residential towers as part of the State residential and commercial renewal planning as illustrated in Plate 1.2. There was scope here for comprehensive value capture for public benefit (Chap. 3).

Looking at transport planning and value capture innovations further, Steve Harrison (2010), Urban Development Institute of Australia (UDIA) Gold Coast President, explains how transport improvement benefits don't always fully flow to the public. He notes that many developers, for example, grab land around planned rail stations for land value gains (to be addressed in Chaps. 3, 4, 5 and 6). More information on value capturing on the Gold Coast can be accessed at www.goldcoast.com.au.

Consult Australia (2015) notes the importance of the Australian States to now consider value capture. In backing up its findings, Consult Australia notes four key subjects in VC review: State Context; Pilot Study Programs; Value Capture Projects; and, Value Capture Legislation. Expanded details on this are contained in Addendum

Plate 1.2 Artist's impression of planned development at future Waterloo train station. (Source NSW Government 2020)

1.2. Finally, Consult Australia concludes from these key subjects that Australia needs to act now on value capture. If not (Consult Australia notes) there will be inadequate funds for needed infrastructure. As a result there could be (Consult Australia notes) a steadily diminishing quality of life in cities for future generations. Finally, Consult Australia notes that a study by KPMG for the Sunshine Coast (Australia) light rail project indicated that a well-designed and articulated value capture strategy could contribute in the order of 10% to 20% of that project's $1.8 billion cost (see #1 in Addendum 1.1). To expand on the subject, different approaches (including projects) to value capture is addressed next.

1.2.4 Value Capture Approaches

Harrison (2010) (noted earlier) comments that the NSW Government has, in a few instances, adopted a value capture approach to public infrastructure provisions. The Northwest Metro rail project as noted earlier is one example (https://www.sydneymetro.info/northwest/project-overview). The government bought extra semi-rural land surrounding the proposed new stations for use during construction, with a view to later selling this land to developers. Harrison also points out that if a government already owns land that will increase in value as a result of the new infrastructure, then the city can capture that value. Most road and transport authorities are established with the power to acquire land for the purposes of building, operating or maintaining public infrastructure. Harrison believes, however, Australian authorities generally don't have power to acquire land for the purposes of later selling it at a profit to help fund the infrastructure. Thus, he concludes that the legislation would need to be amended to give authorities these powers (or a new authority established with these powers).

Looking further, Harrison (2010) notes that there is a wide expanse of value capture potential. He introduces associated development opportunities as part of infrastructure projects (say within a project tender). He states that this approach was adopted for the Chatswood Transport Interchange in Sydney (noted earlier). Here the Private, Public, Partnerships (PPP) tender contract included the right to develop a shopping centre and residential towers above the new railway station and transport interchange. Harrison, however, cautions governments to ensure that a contract has leverage on the private sector to develop the public infrastructure.

Moving further, Dr. Lindsay Taylor (2016) sees Voluntary Planning Agreements (VPAs) (introduced earlier) related to value capture. Taylor notes that Section 93F of the NSW Environmental *Planning and Assessment Act 1979* (EPA Act) establishes a statutory system of bargaining for community benefits (engaging planning authorities, developers and land owners). Here increased development potential relating to land is exchanged for community benefits funded by the increased land value. He notes that the instrument is contained within a VPA. Taylor says these VPAs have a potentially broad field of operation, sufficient among other things to accommodate the practice of *value capture*. He notes that this is similar to what occurs in connection with *incentive zoning* in the United States and *s106 Agreements* in the United Kingdom.

Commenting on the Australian directions in value capture, Harrison (2010) notes that the Federal government needs to explore value capture as a way of funding major federal infrastructure projects in the future. The Australian House of Representatives Standing Committee on Infrastructure, Transport and Cities (https://www.aph.gov.au/search/url/Committee/24099_0*)* completed an inquiry on the related issue of transport connectivity and funding (#1.2 in Addendum 1.1). This inquiry had a particular focus on how value capture mechanisms could be used to deliver transport projects.

Harrison also notes the recently adopted NSW government's *Special Infrastructure Contribution* (SIC) (2016), a variant of value capture as a one off payment (essentially a levy on developers). The SIC will be applied, for example, to the current Parramatta light rail corridor project (to open in 2022) (Plate 1.3). The State has adopted a SIC levy on developments along the corridor. The SIC levy is $200 per square metre of gross floor area of new residential developments in defined districts near new heavy and light rail stations (addressed in coming chapters). As of late 2020, however, application of the SIC to this corridor had not been activated (stalling at the State government level).

This SIC levy could be used, for example, to partially fund the construction of 10,000 new homes and a railway station (underway) at Waterloo (Inner Sydney) (examined in Chap. 3). Moving to the Federal government, it is also looking at ways to use value capture to fund a rail link to the proposed Western Sydney Airport near Liverpool. This would include giving developers the rights to build an underground

Plate 1.3 Proposed Parramatta Light Rail. (Source NSW Government 2020)

station and developing surrounding land for commercial and residential purposes. Given variations of value capture programs in different countries (to learn from), this subject is examined next.

1.2.5 Value Capture in Different Countries

Looking at value capture beyond Australia, the Lincoln Institute's Land Policy's Conference (2015) (in Brazil) featured urban renewal and value capture. Value capture initiatives reviewed at the conference included those in Chile, Panama and Brazil. Likewise, the Washington DC based Urban Land Institute illustrates the strength of value capture in its published European guide to value capture finance (locate via http://uli.org). Turning to Australia and noting the above, Consult Australia (2015) comments that an Australian value capture program would need to respond to a number of items, including: Commonwealth; state and local procurement; taxation; land use; town planning; and, other relevant policies and legislation. Consult Australia also notes the recent study on the subject by the Australian Bureau of Infrastructure, Transport and Regional Economics (BITRE) (investment.infrastructure.gov.au/files/value_capture/Value-Capture-Discussion-Paper.pdf).

The Bureau (#2 in Addendum 1.1) estimates that the public's investment in 128 road and rail projects in Australia returned (public benefits) $2.65 for every $1 invested. This investment has a present value share of net public benefits of $62 billion. In the UK, London's Crossrail project as noted earlier (an eight station, 21 km addition to the metropolitan area's underground commuter rail network) (#3 in Addendum 1.1) is expected to return (cost-benefit) between 3 and 5 pounds for every pound invested.

Transport and housing projects are often combined with value capture programs. Thousands of homes, for example, and a shopping centre and hotel were built around or near Tung Chung Station (Hong Kong). This was part of Mass Transit Rail's (MTR) rail plus property model. For instance, MTR's first international development project was in Shenzhen, China. Called Tiara (Plate 1.4) (#4 in Addendum 1.1), the project is also built around a train depot and includes about 1700 resident apartments.

In many countries value capture revenues have been used to pay for a variety of costs in urban renewal sites (in addition to transport infrastructure). Governments have used these value capture revenues for: property acquisition; rehabilitation and renovation of historic structures; and, construction of new or improvements to existing civil infrastructure. New revenues from public investment sites often extend beyond their immediate locations. This can include, for example, the wider community benefits of light rail or road improvements. Examining planning in Chicago (where value capture districts exist), some renewal programs integrate land uses (for example housing and transport). Consult Australia (2015) notes a 2006 study was completed on 89 Chicago value capture districts (#5 in Addendum 1.1) (spread across 67 municipalities). This study found that mean annualised property values in

Plate 1.4 Mass Transit Rail (MTR) development project, Hong Kong. *Left to right*: Train depot with developments; and, residential high rise buildings. (Source MTR 2020)

these districts (given investments) increased by 35 per cent between 1983 and 1993. This compared with a 6 per cent increase in overall municipal property values. Urban renewal programs thus not only revitalise cities but also enhance public tax revenue.

Value can also be created by government decisions to rezone land (i.e. housing or public transport). This means land can be used for higher value land uses. By planning and timing such decisions with other value capture mechanisms, governments can capture a greater share of the value it creates. Harrison comments that the Hong Kong Mass Transit Rail (MTR) (noted earlier) improves the land value with a series of linked rail infrastructure projects (i.e. rail stations, interchanges and residential and commercial developments). Finally, he notes, the Hong Kong model has been deployed with such success that the government is able to fully cover the cost of the mass transit system without any government subsidy. Harrison cautions, however, that the density and associated land values of Hong Kong means that it is unlikely the model could be deployed at low densities with the same degree of success as above. He states, however, there are dense urban locations in Australia where transport infrastructure is presently inadequate and where a MTR-style model could deliver good results given the range of value capture programs available.

Governments can also capture property value increases resulting from development of their new infrastructure. A city can, for example, purchase land that will increase in value as a result of a proposed project prior to announcing the project. The government can then sell the land after its value has increased (noted earlier). World cities continue to expand value capturing initiatives. Consult Australia notes, for example, that The Denver Union Station (State of Colorado) (#6 in Addendum 1.1) (hub for Denver's light rail network) redevelopment project captured $135 million of its $446 million cost through value capture (or roughly 30% of the project's capital cost). Consult Australia also notes that Hong Kong's Mass Transit Rail (MTR) (noted above) ((#7 in Addendum 1.1) system is fully funded by property development gains. In this example, MTR acquires land for stations at values based on a 'no-rail scenario' and improves the land with infrastructure, including transit stations. Development sites around the stations are then leased at higher values with the infrastructure in place. Using a fully integrated transport and property development model reflects an expanded application of value capture in a densely developed urban environment, Hong Kong's public transport system thus pays for itself. The company's development around the Tung Chung Station (close to the Hong Kong airport) (#8 in Addendum 1.1) has 38 high rise towers with more than 12,000 flats, 97 houses and a hotel and big shopping centre. Harrison, however, puts forward the question 'is the MTR approach migrating to Australia and does Australia want it'?

Looking at Paris (#9 in Addendum 1.1), its metro system is undergoing a major 200 km extension costing €30 billion. It's understood the vast majority of funding will be from a regional tax on commercial buildings that benefitted directly from the project. Other value capture international projects are listed in Addendum 1.1, including: Emirates (#10) (stadium and community facilities); Barcelona (#11) (private development locks and community orientated facilities); Copenhagen (#12) (Metro rail and facilities); HafenCity, Hamburg (#13) (inner city urban development); and, City of Berlin (#14) (arena and facility planning). Given these above examples from around the world, it is concluded that value capture's contributions to urban planning cover a wide range. Contrasting views on value capture is examined next.

1.2.6 Contrasting Views on Value Capture

There is a wide cross section of different views on value capture. Newman (McIntosh and Newman 2017) takes a more comprehensive view to LVC. He writes that just putting levies on developers was treating the levy as isolated and standing alone. He indicates that a more radical MTR-style (noted above) development-focused approach to public transport and housing was needed. He notes that if you want authorities to get levy money from the private sector, authorities have to run mass transport projects as more comprehensive redevelopment projects (rather than solely as transport projects). In a funding model therefore, private sector consortiums would be invited to bid for rights to develop. This can include, for example, a rail system and the surrounding residential or commercial sites, aiming at the lowest expense to the government. This is partly the intention of the NSW government at the Waterloo renewal site (noted earlier and addressed in Chap. 3).

Not everyone however is sold on the value capture concept. Chris Johnson (2018), as the chief executive of developer lobby group Urban Taskforce, says he is concerned value capture would become a "dangerous buzz word" without substance of testing. Johnson's main worry is that residential property developers will pass on any levies to buyers by boosting house prices. Johnson commented that "the government needs to be aware that value capture will increase the price of housing and be inflationary". He notes that Urban Taskforce is not against the concept of value capture per se. While developers can also profit with value capture, Johnson is worried the floor space linked conditions could inhibit investments (especially medium sized investors). He notes that there seems to be a threshold reached in terms of selling prices in Sydney.

The danger of adding a high contribution amount for value capture is that development may not be economically viable. Thus Johnson argues that funds will not then flow to the infrastructure. Johnson favours further discussion on the subject between government and industry (*Source* NSW Government News 2017). He also indicates that developers are concerned that the cost of new housing will increase if developers have to fund rail infrastructure and social housing (noted earlier outlining value capture projects). One concern expressed, for example, is that the NSW government will want private developers to fund the rebuilding of the 2000 existing social housing units as part of the urban renewal of the Waterloo Estate. He states that the Government has required the private sector to fund new social housing in urban renewal developments at the Riverwood and Ivanhoe Estates (subsidizing this through new private housing approvals). The New South Wales Government's Draft Redfern Waterloo Built Environment Plan (2011) calls for a mix of private and social housing. The plan (Johnson notes) implies that the private development will subsidize the social housing (examined further in Chap. 3).

Moving to coordination of value capture, SGC (2016) in *Report on Value Capturing* notes some dangers here with LVC. SGC states that if government does not properly coordinate LVC the program it can have an adverse effect on development and government. SCG notes that this can affect housing affordability, as well as raising costs of development. Co-ordination and a clarity in LVC programs (SGC argues) are required in balancing VC with other planning policy concerns.

Moving briefly from housing and transport to environment (i.e. open spaces), research has also quantified land value gains from open space improvements. This is addressed in *Measuring the Economic Value of a City Park System* (Harnik and Welle 2009) (http://cloud.tpl.org/pubs/ccpe-econvalueparks-rpt.pdf). There is a conservative value of 5 percent increase added to values of dwellings within 500 feet of these parks. Other studies have revealed that well designed environment areas and parks tend to add 15 percent to the value of a proximate dwelling. There are however value capture variations and confusions, as examined next.

1.2.7 Variations and Confusions About Value Capture

There are variations to and confusions about the benefits of value capture, as Harrison (2010) explores. He explains, for example, how voluntary contributions can be made to authorities. These contributions can be in lieu of value capture provisions by a developer. He also states that many organizations (besides developers) benefit from the provision of public infrastructure. He then notes that these organizations (i.e. commercial companies) may be prepared to voluntarily contribute to the funding of public infrastructure. In instances this contribution may speed up the completion of an infrastructure project. The contribution may also allow a company to obtain greater input into decisions on how value capture is progressed. Harrison gives the example of a private organization making a contribution for a specific road upgrade to accommodate its business operations. He argues there thus should be an opportunity for the company to invest directly in public road infrastructure projects.

There is, however, at times confusion about what value capture actually is. The NSW State government, for example, is facing escalating community concerns about plans for value capture (especially the State's redevelopment plans at different locations). Urban Growth (under the State at the time and in 2019 was merged into Infrastructure NSW) (infrastructure.nsw.gov.au), for example, pursued new infill urban densification. Communities, however, are deeply worried that established residential communities in older inner suburbs will be overwhelmed by new developments (i.e. often motivated by apartment block demands). Many sections of inner Sydney communities remain confused about value capture and betterment levies. Citizens thus overall want more information on value capture, especially in relation to transport infrastructure and urban renewal projects (examined in Chap. 2).

To further address value capture questions (in addition to Addendum 1.2 referred to earlier), key subject documents (complementing those noted above) can be accessed at two Prosper Australia web sites. These are noted below (second site gives a historical perspective of value capture within Australia).

https://www.prosper.org.au/26p/
http://www.prosper.org.au/land-value-capture/value-capture-a-historical-perspective/

Finally, other resources that outline views on value capturing and its importance of value capture across a number of land uses are listed in Addendum 1.3. Value capture programs are thus examined next.

1.3 Conclusions

There are a number of conclusions reached in canvassing urban planning in the post coronavirus era and the future role of value capture planning (VCP). Firstly, there is likely to be significant changes to the design of cities in the post coronavirus era (i.e. housing; public and open spaces; and, sustainable transport). Secondly, in addressing future renewal of cities, there is potential scope for the adoption and application of VCP (applicable to any country and as illustrated in case city examples). In addition, it was postulated how VCP could be used as a tool in achieving equitable planning outcomes (e.g. housing, open spaces and transport) as examined in coming chapters.

1.4 Summary

The chapter outlined planning for renewing cities and regions using value capture (VC). Firstly, the implications of the coronavirus pandemic (2020) for future urban planning (including value capture) were reviewed (i.e. housing, open spaces, and transport). The focus of the chapter then moved to cities adopting VC policies to guide renewal and development. Those governments that use VC within their planning were then examined. It was explained how VC policies are triggered by government decisions. These included: land up-zoning and designation of major government infrastructure projects. The chapter then addressed key aspects of VC planning. Finally, the chapter explained how value capture is administered through value capture planning (VCP) programs, including engaging the governing body, developers and the community.

Addendum 1.1: Table of Value Capture Projects – International Examples

This table was completed (Rauscher) to illustrate the breath and variety of value capture project applications around the world. They are listed by location and project, including value capture details. By project types (a-z and numbers of projects listed) there are: commercial and residential (1); heavy rail (and related developments) (5); land use rezonings (4); light rail (2); and, sports infrastructure (2). Note the high number of transport projects (7 in all out of the 14 projects) using value capture.

Project	Location	Project type	Outline of project	Value capture details
1	Sunshine coast, Australia	Light rail project	A $1.8b new light rail	Value capture income contributed bewteen10–20% of project cost
2	Australia wide	Roads and rail	Total of 128 road and rail projects across Australia	Public benefits estimated at $2.65 for every $1 invested (net public benefits of $62 billion)
3.	London	Crossrail	Eight station, 21 km addition to rail network	Cost-benefit between 3 and 5 pounds for every pound invested.
4.	Shenzhen, China	MTR Tiara Project	Built around a train depot and includes about 1700 resident apartments	Community projects built at no public cost
5.	Chicago	Renewal and Land Uses	Renewal programs and integrating land uses (i.e. housing and transport).	89 Chicago value capture districts within 67 municipalities
6.	Denver	Denver Union Station	State of Colorado hub for Denver's light rail network	Value captured at $135 million (30%) of $446 million station cost

(continued)

Project	Location	Project type	Outline of project	Value capture details
7.	Hong Kong	Mass Transit Rail (MTR))	MTR has built an extensive rail system engaging property developments	MTR System is fully funded by property development gains
8.	Hong Kong, Tung Chung	Rail Stations	Tung Chung Station has 38 high rise towers (12,000 flats) 97 houses and a hotel and shopping centre	New rail stations have been built under value capture encompassing residential and other land uses
9.	Paris	Metro Rail	Metro system included a 200 km extension costing €30 billion	The majority of funding is from a regional tax on commercial buildings that value benefitted directly from the project.
10.	Emirates	Emirates Stadium	New stadium and community facilities	Value capture of social housing, new park, day nurseries and education centres were community improvements
11.	Barcelona, Spain	22@Barcelona,	Land-use designation of 115 privately-owned old blocks	Planning changes dramatically increased the land's potential value to private owners. Result was value capture of the community-orientated improvements (estimate EUR 1 billion)
12.	Copenhagen, Denmark	Ørestad Development Scheme	The increased accessibility to the adjacent land raised demand for land amongst developers and investors	Captured values paid for the construction of the Metro (EUR 2.3 billion).
13.	HafenCity, Hamburg, Germany	Europe's Largest Inner City Urban Development Zone	5,500 new housing units, more than 40,000 jobs and 10 kilometres of quayside promenades	The holistic masterplan for the area created a Vision. This vision raised demand for sites in the area (resulting in value capture benefits).
14.	City of Berlin	Multifunctional Facility Planning	Arena and ½ million sm gross floor area of mixed use	City used a number of value capture contributions from the developer for local public infrastructure

Source Table constructed by Ray Rauscher (2020) from various sources

Addendum 1.2: Land Value Capture – Adoption and Implementation

Consult Australia (2015), in explaining value capture in *Commenting on Land Value Capture Policies for Australia,* listed four key subjects. These are: (1). State Context; (2). Pilot Study Programs; and, (3). Value Capture Projects; and, (4). Value Capture Legislation. Each of these subjects is summarized below by Consult Australia.

State Context

States and territories throughout Australia are ramping up infrastructure investment to build new infrastructure, maintain and repair deteriorating assets, support population and employment growth, and boost economic performance. Aspects of land value capture include, as follows: precinct based planning; smart growth principles; value capture programs; and, urban renewal efforts.

Precinct Based Planning

Precinct-based planning and funding allows stakeholders to focus on specific infrastructure needs, develop tailored solutions, clearly identify beneficiaries and better define funding options. It also allows sponsors and program managers to track results earlier and make program improvements. Pilot projects with well-conceived funding guidelines and objectives are excellent means of testing and improving concepts and techniques before full program rollout occurs. The location and boundaries of value capture improvement precincts should be carefully selected to include existing and potentially new complementary commercial and public activities and investments that can be further levered by the primary infrastructure investments.

Smart Growth Principles

Smart growth principles provide excellent planning frameworks to leverage public sector investments in our cities. The State of New South Wales (NSW) Department of Planning and Environment (DPE) and UrbanGrowth NSW employ precinct-based planning frameworks, including the DPE's Urban Activation program, Priority Precincts program and Local Infrastructure Growth Scheme.

Value Capture Programs

Value capture programs incorporate incentives within local government planning controls to attract private investment, promote economic development and capture revenue for infrastructure. Complementary programs developed with urban renewal and other agencies can also achieve social outcomes, such as affordable housing.

Urban Renewal Efforts

Planning controls and compulsory acquisition powers need to provide urban renewal authorities with the ability to undertake timely and sometimes widespread urban renewal programs where necessary or desirable in the broader public interest. It is recommended that Commonwealth (federal) and state governments collaborate to develop a common framework and commission pilot projects to assess the wider economic benefits of infrastructure and urban renewal projects.

Pilot Study Programs

Pilot projects should be undertaken in partnership with state and local government agencies, professional associations, research institutions and the private sector. These bodies could contribute to the development of the legislative and financial arrangements needed to harness these benefits. It is proposed that the Commonwealth and state government undertake practical research into value capture methods as a funding supplement for state and local infrastructure projects. This could be accomplished by establishing pilot study programs to be undertaken by local government councils, urban renewal authorities and other state agencies to fully explore the opportunities and obstacles to this funding method.

Transport improvement precincts around metro stations typically extend between 800 and 1000 metres from the station. Numerous studies have demonstrated that the measureable benefits of metro stations are generally contained within this distance. Downtown improvement districts in regional cities and small towns in North America are typically contained within several blocks of the commercial core. Each location will have different needs and characteristics and must be individually determined.

Value Capture Projects

Value capture methods require detailed descriptions and financial models to be prepared for "Base Case". Scenario testing of these alternatives as a minimum is necessary to demonstrate the impact of the value capture intervention programs on a variety of financial and built environment parameters. Project scenarios evaluate the proposed interventions of the value capture program for comparison against the Base Case, including:

3.1 Changes to zoning and development controls made possible by project investments, such as additional public transport capacity enabling higher density development
3.2 Increases in public revenues from increased retail and business activity, residential and commercial property development, and employment growth
3.3 Improvements to the public domain designed to attract pedestrian activity and access, such as commercial centre revitalisation programs since urban renewal and transport improvement programs can take many years to reach their full potential. Scenarios should extend for as long as necessary to fully capture the programs benefits. For example, value capture programs in North America typically forecast expenses and revenues at least 20 years into the future.

Value Capture Legislation

Limited value capture programs can be developed under existing legislation in many jurisdictions. In NSW for example, voluntary planning agreements (VPA) are being used to accept payments from developers for additional floor space in some local government areas.

Addendum 1.3: Land Value Capture Resources

Selected Books and Articles

Batt, H. William (Jan 2001) Value Capture as a Policy Tool in Transportation Economics, American Journal of Economics and Sociology

Levinson, David M. and Emilia Istrate (2011) Access for Value: Financing Transportation Through Land Value Capture. Brookings

Royal Institution of Chartered Surveyors (RICS) and the Office of the Deputy Prime Minister (ODPM) (Oct 2002). *Land Value and Public Transport*, London

Scottish Executive (2004) Developing a Methodology to Capture Land Value Uplift Around Transport Facilities: Executive Summary. Scotland

SGS Economics and Planning (2016) (Fox/Stapledon) Report on Value Capturing. Melbourne

Smith, Jeffery J. and Thomas A. Gihring Nov 2004) Financing Transit Systems Through Value Capture: An Annotated Bibliography, in Victoria Transport Policy Institute Encyclopaedia, Victoria Canada

Selected Web Sites

In addition to writing there are a number of key web sites on the subject of land value capture, including:

Centre for Land Policy Studies (UK)
Earth Rights Institute's Land Rights and Land Value Capture Course
Henry George Foundation
Land Values Research Group (Australia)
Lincoln Institute of Land Policy (USA)
The Center for the Study of Economics

References

AEC (2016) Wickham masterplan economic and market analysis. AEC, Newcastle

Bass RM (2020). https://www.brookings.edu/blog/the-avenue/2020/03/25/the-qualities-that-imperil-urban-places-during-covid-19-are-also-the-keys-to-recovery/. Brookings Institute, New York

Buxton M (ed) (2016) Planning Melbourne – lessons for a sustainable city. CSIRO, Canberra

Chen E (2020) Collectivity and the common good: how housing will change thanks to Coronavirus. Arch Daily's, Brazil

City of Sydney (2012) Development guidelines – providing community infrastructure in Green Square. City of Sydney Council, Sydney

Clare K (2016) Sydney and high-rise (housing futures conference). Architecture Media, Sydney

Club of Rome (2019) The planetary emergency plan, UN, New York

Consult Australia (2013) Capturing value. AECOM Consult Australia and Sinclair Knight Merz, Sydney

Consult Australia (2015) Commenting on land value capture policies for Australia. AECOM Consult, Sydney

Consult Australia (2016) Land value capture roadmap. AECOM Consult, Sydney

Department of Planning, Industry and Environment (2000) Planning and assessment regulation 2000. NSW Government, Sydney

Department of Planning, Industry and Environment (2016) Practice note and draft planning circular on development agreements and funding infrastructure. NSW Government, Sydney

Forsyth A (2020) What role do planning and design play in a pandemic? Harvard Univ. Graduate School of Design, Harvard

Harnik and Welle (2009) Measuring the economic value of a city park system, Chicago. The Trust for Public Land

Harrison (2010). www.goldcoast.com.au 26/9/2010, Australia, Gold Coast

Jacobs J (1961) The death and life of great American cities. Blackwell's, New York

Johnson C (2018) Parramatta value capture will stop development unless industry input is considered. Property Observer Newsletter 6 Sept 2018, Sydney

Karlen A (1995) Man and microbes: disease and plagues in history and modern times. Touchstone, New York

Klaus I (2020) Coronavirus and urban resilience. City Lab, Ridgefield

Lucas C (2016) *Review of high-rise building*. The Age, Melbourne

Mao J (2015) The airborne transmission of infection between flats in high-rise residential buildings. Build Environ Issue 94:516–531. Science Direct, Indiana

McIntosh N et al (2017) Framework for land value capture from investments in transit in car-dependent cities. J Transp Land Use 10(1) Perth

NSW Government (1979) NSW environment, planning and assessment act (1979). NSW Government, Sydney

Panczak R (2013) High life in the sky? Mortality by floor of residence in Switzerland. Eur J Epidemiol 28:453–462. Springer, Switzerland

Planetizen (2020) Debating future cities and urban density after the Pandemic. Blog Post https://www.planetizen.com/blogs/108814-debating-future-cities-and-urban-density-after-pandemic

Rauscher RC, Momtaz S (2014) Sustainable communities: a framework for planning. Springer, Cham

Rauscher RC, Momtaz S (2015) Brooklyn's Bushwick – urban renewal in New York, USA. Springer, Cham

Rauscher RC (2018) New York neighborhoods – addressing sustainable city principles. Springer, Cham

Rice K (2020) What our cities will look lake after Coronavirus. Urban Edge, USA

Rosmarin (2020) Coronavirus and public spaces. University of Melbourne, Melbourne

SGC (2016) Report on value capturing. SGC, Melbourne

Surkes S (2020) Coronaviruses expected to increase with global warming. Times of Israel, Web Ref. https://www.timesofisrael.com/viruses-expected-to-increase-with-global-warming-expert/ Tel Aviv, Israel

Taylor L (2016) (Source Lindsay Taylor web 18 May 2016), Sydney

World Health Organisation (WHO) (2020) Global change summary https://www.who.int/global-change/summary/en/index5.html

Chapter 2
Greater Sydney and Regions' Renewal – Value Capture Planning Model

Abstract This chapter outlines the development of a value capture planning (VCP) model and details how it will be applied to study areas (4 cities within urban corridors in Australia). Firstly, the potential opportunities for the application of value capture to the renewal of cities is outlined. Attention is drawn especially to the implications of the coronavirus pandemic (2020) on future urban planning (Chap. 1). The VCP model is then outlined and selected planning principles of the model explained. These principles and components (in brackets) are: Equitable *Housing* (affordable, social and market housing); *Public and Open Spaces* (natural areas, open spaces and public spaces); and, *Sustainable Transport* (rail, bus, and active transport). The chapter then explains the application of these planning principles and the developer provision and/or levy alternative. The concept of extending the VCP model application from a precinct level (i.e. renewal areas) to areas beyond the precinct is also commented on.

The Greater Sydney and region centres planning is then introduced. Study areas and case study renewal precincts within these areas are selected for model application. Comments are also offered on surveys to be completed within the selected renewal areas. These areas include by type of urban area (renewal area and city in bracket): (1) an inner city area (Central to Eveleigh Renewal Area (CERA), City of Sydney); (2) a middle ring city (Sydenham to Bankstown Urban Renewal Corridor (SBURC), City of Canterbury Bankstown); (3) a growth centre city (Gosford City

---

Dedication There are many people in the history of cities and renewal who have left major legacies. This chapter is dedicated to two of them (Plate 2.1), Florence Mary Taylor (architect and engineer) and Neville Wran (NSW Premiere).

Florence Mary Taylor CBE (1879–1969) was the first qualified female architect and the first woman to train as an engineer in Australia. She was also the first woman in Australia to fly in a heavier-than-air craft in 1909 and the first female member of the UK's Institution of Structural Engineers in 1926. However, she is best known for her role as publisher, editor and writer for the influential building industry trade journals (*Source* Wikipedia 2020).

Neville Kenneth Wran, AC, CNZM, QC (1926–2014) was an Australian politician who was the Premier of New South Wales from 1976 to 1986. He was the national president of the Australian Labor Party (ALP) from 1980 to 1986 and chairman of both the Lionel Murphy Foundation and the Commonwealth Scientific and Industrial Research Organisation (CSIRO) from 1986 to 1991 (*Source* Wikipedia 2020).

R. Rauscher, *Renewing Cities with Value Capture Planning*,
https://doi.org/10.1007/978-3-030-62958-8_2

Plate 2.1 Dedications: Florence Mary Taylor and Neville Wran

Centre Revitalisation (GCCR), Central Coast Council); and, (4) a regional capital city (Newcastle City Centre Renewal Area (NCCRA), City of Newcastle).). Finally, case study precincts within each study area city are selected and comment offered on surveys to be undertaken for each precinct.

2.1 Developing a Value Capture Planning (VCP) Model

Having examined the background on urban renewal and value capture (Chap. 1), the value capture planning (VCP) model is outlined here for any local, state or federal government to consider in their urban planning. Developing a VCP model needs to accommodate locality factors such as: an authority and state's current plans; identification of planning precincts; and, projection of key potential development and renewal sites within these precincts (see also Chap. 1 Addendum 1.1 Land Value Capture – Adoption and Implementation). The proposed VCP planning principles (components in brackets) include: *Equitable Housing* (affordable, social and market housing); *Public and Open Spaces* (natural areas, open spaces and public spaces); and, *Sustainable Transport* (rail, bus and active transport). The second stage of the VCP model is *Developer Provision and/or Levy Alternative*. Each of these VCP planning principles and the Developer Provision and/or Levy Alternative are now addressed, commencing with equitable housing.

2.2 Equitable Housing

Housing in most cities is the most important provision of population need, after food. Equitable housing signifies a balance of housing types, including: affordable, social and, market housing. The first Table (2.1) (Equitable Housing) incorporates four key steps of value capture (VC) for an authority to apply. An example of planning new housing at 4000 units is taken within the table.

Application Summary
A checklist is provided to identify housing types (Step 1). Step 2 is meeting criteria of need for housing types. Step 3 is addressing design guidelines for housing types. The projected units of new housing by types (Step 4) can be calculated by number and %. The planning body can then use these numbers within developer agreements (Table 2.4).

2.3 Public and Open Spaces

Public and open space provisions in most cities is a critical principle of planning for a population. Components include: natural areas, open spaces, and public spaces. The second VCP Table (2.2) (Public and Open Spaces) incorporates five key steps.

Application
A checklist is provided to identify public and open spaces types (Step 1). Step 2 is meeting criteria of need for public and open spaces provision. Step 3 is projecting new public and open spaces by type. Step 4 is addressing design guidelines for

Table 2.1 Value Capture Planning (VCP) table: equitable housing

Step	1	2	3	4
Checklist	Housing types	Meet criteria of need for housing types (examples)	Addressing design guidelines for housing types (examples)	Projected # and % of new housing units(example)
	All new housing	Housing standards	Variety of designs	4000 (100%)
y	1. Affordable housing	Special housing groups	Innovative architecture	1200 (30%)
y	2. Social housing	Social support needs	Maintenance efficiencies	800 (20%)
y	3. Market housing	Housing as income compatible	Range of housing style choice	2000 (50%)

Steps as per columns	
1	Tick the checklist for housing types that apply
2	Adopt criteria for housing provision
3	Refer to design guidelines for housing
4	Projected # and % of new housing units by type

Table 2.2 Value Capture Planning (VCP) table: public and open spaces

Step	1	2	3	4	5
Checklist	Public and open space types	Meeting criteria for public and open spaces provision (examples)	Projected new public and open spaces	Address design guidelines for public and open spaces (examples)	Cost of public and open spaces provision by type
✓	1. Natural areas	Natural area regeneration	ha	Amount of regeneration	$
✓	2. Open spaces	Active and passive spaces	ha	Connecting public and open spaces	$
✓	3..Public areas	Need for public spaces	ha	Amount of spaces required	$
	Totals		ha		X $

Steps as per columns

1	Tick the checklist for public and open spaces types that apply
2	Adopt criteria for public and open spaces provision
3	Projected new public and open spaces
4	Refer to design guidelines for public and open spaces
5	Cost of open spaces provision by type (total cost shown as X$)

public and open spaces. Step 5 calculates the cost of the new public and open spaces. The user of the table can then use these numbers in formulating developer provisions and/or levy alternatives among stakeholders (Table 2.4).

2.4 Sustainable Transport

Sustainable transport is the third planning principle of the VCP model and an important infrastructure provision for the planning of communities. Sustainable transport components include: rail (heavy and light); bus; and active transport. This third VCP Table (2.3) (sustainable transport) incorporates five key steps.

Application

A checklist is provided to identify sustainable transport provisions (Step 1). Step 2 is meeting criteria for the transport provisions. Step 3 is addressing design guidelines for sustainable transport. Step 4 is projecting sustainable transport provisions new public and open spaces by type. Step 5 calculates the cost of new sustainable transport provisions. The user of the table can then use these numbers in formulating developer provisions and/or levy alternatives among stakeholders (Table 2.4).

Table 2.3 Value Capture Planning (VCP) table: sustainable transport provisions

Step	1	2	3	4	5
Checklist	Sustainable transport provisions	Meeting criteria for sustainable transport provisions (examples)	Address design guidelines for sustainable transport (examples)	Projected new transport provisions (i.e. infrastructure)	Cost of new transport provisions by type
✓	1. Rail (heavy and light)	Demand and access	New station, transport or interchange	Type provision	$
✓	2. Bus	Demand and access	Bus transit way	Type provision	$
✓	3. Active transport	Cycleways, pedestrian paths, and share ways	Cycling and walking links	Type provision	$
	Total				Y $

Steps as per columns	
1	Tick the checklist for sustainable transport types that apply
2	Adopt criteria for sustainable transport provisions
3	Refer to design guidelines for sustainable transport provisions
4	Projected new transport provisions
5	Cost of new transport provisions by type (total cost shown as Y$)

Table 2.4 Value Capture Planning (VCP) table: developer provision and/or Levy alternative for public and open spaces, and transport infrastructure

Step	1	2	
		Developer provision and/or levy alternative per new housing unit	
Check list	Projected units of new housing by type and % (see Table 2.1)	Public and open spaces X $ (total) (see Table 2.2)	Transport provision Y $ (total) (see Table 2.3)
✓	All new housing (example) 4000 (100%)	**X** $ + Y $ ÷ 4000 units = Z $ per unit	
✓	1. Affordable housing 800 (20%)	$ per unit for 800 units	
✓	2. Social housing 1200 (30%)	$ per unit for 1200 units	
✓	3. Market housing 2000 (50%)	$ per unit for 2000 units	

Steps as per columns	
1	Tick the checklist for housing types that apply and projected new housing by type and %
2	Value capture levy per new housing unit for public and open spaces, and transport infrastructure (Z $ per unit)

2.5 Developer Provision and/or Levy Alternative

The second stage of the VCP model is the Developer Provision and/or Levy Alternative. Developer provisions (i.e. within development agreements) of value capture planning (VCP) allows the developer a choice of providing a facility and/or (if more appropriate) a levy alternative. The forth VCP Table (2.4) incorporates two key steps.

Application
A checklist is provided to address the projected new housing (Step 1) by type and % from Table 2.1. Step 2 identifies the developer provision or levy alternative per new housing unit for public and open spaces (Table 2.2) and sustainable transport provisions (Table 2.3). The user of the table can then use these numbers in formulating developer provisions and/or levy alternatives among stakeholders.

2.6 Extending the VCP Model Beyond the Precinct

The concept of extending the VCP model beyond the precinct to other areas is addressed. While the model is formulated to apply to areas such as renewal zones the application to other areas (i.e. urban corridors or whole of cities) is addressed in each chapter. Greater Sydney and regional centres bordering Greater Sydney are introduced and examined next. From this examination four case study areas will be selected to apply the VCP model to in subsequent chapters.

2.7 Greater Sydney Introduction

The shaping of any city, including the study area of Greater Sydney and Regions, is not always a smooth path. Cities are faced with social, economic and environmental challenges related to development. Here (Plate 2.2) is Greater Sydney by localities, providing a glimpse of the spread of the city (Pacific Ocean to the right). Here the city is ever expanding outwards from Central Sydney (at water's edge to the right-middle in purple, just above Eastern Suburbs in pink and South Sydney in Green).

Most cities today (as is the case of the cities with Greater Sydney) accept the principle of ecological sustainable development (ESD). In addition these cities often adopt indicators of sustainability (Rauscher 2014). Principles of the United Nations' Agenda 21 and Local Agenda 21 are also normally acknowledged. There are usually hurdles to jump in cities moving to adopt sustainable urban planning (SUP) approaches, including within Greater Sydney. One example of divergences of opinion on city development and sustainability within Greater Sydney was during the early 1970s within the City of Sydney. It was here that the phrase 'green bans' was born. The building workers at the time started to look at their engagement to demolish parts of neighbourhoods for new (usually high rise) development. On specific redevelopment sites (from Hunter's Hill (first battle) to The Rocks, Woolloomooloo and Waterloo) building workers concluded (through union representation) that there must be alternatives to construction that compromised community, housing and heritage values. Swaths of the City of Sydney that were

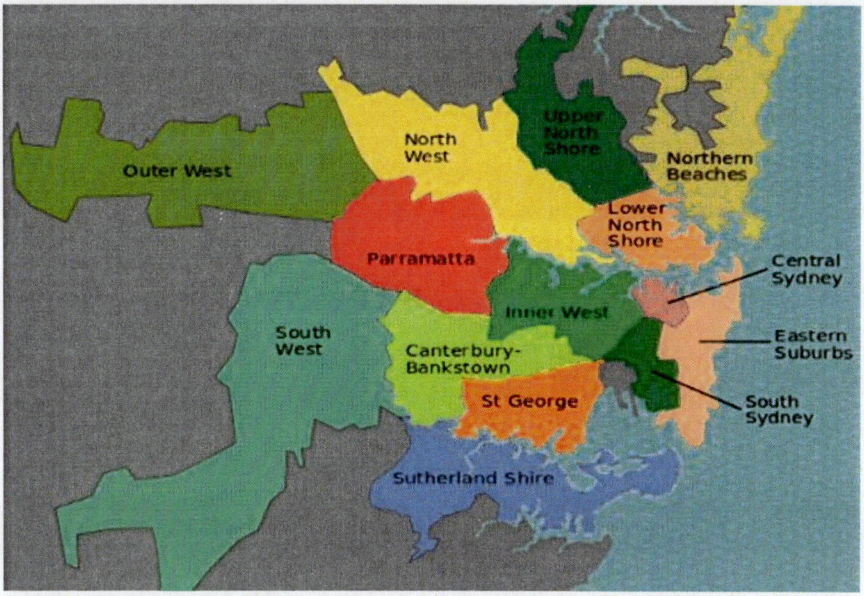

Plate 2.2 Localities of Greater Sydney (Source Dept. of Planning Industry and Environment 2020)

proposed for clearance led to standoff battles. In the end (via negotiations) parties (community, government and developers) came to agreements. Nearly all the green ban actions were effective in preserving sections of neighbourhoods and heritage sites (including bushland at Hunters Hill).

Major sections of inner Sydney today stand as a memorial to the era of the 'green bans'. Plate 2.3 gives a glance of the green bans activity at the time. On the left is union leader Jack Mundey (addressing the workers who had downed tools in a green ban action). The subject of the painting here (centre frame) is Tom Uren, Federal Minister for Urban and Regional Affairs under the Whitlam government (1972–1975). Uren was instrumental in advising the Federal government to get engaged in rescuing several inner Sydney areas (i.e. The Rocks and Woolloomooloo) from redevelopment. The final frame is of a union leader (Joe Owens) explaining to NSW police why a contentious building site was under a green ban worker picket action.

A more recent event likely to affect the future of planning cities in Australia (and the world cities) is the 2020 pandemic of coronavirus (Covid–19). A summary of urban planning implications of this virus (Chap. 1) concluded Greater Sydney and Regions' urban planning will need to be reviewed. One implication noted in is, for example, is a needed review of urban densities and government (and private company) financing of housing, urban infrastructure, public and open spaces, and transport. Planning mechanisms, therefore (such as value capture) could play a part in renewing cities (post coronavirus era and well beyond). Greater Sydney and regional centres planning are addressed next.

Plate 2.3 Green Bans Era in Sydney 1970s**.** *Left to right***:** Jack Mundey addressing workers; Tom Uren painting; and, Joe Owens engaging with Police at picket site. (Source NSW Archives 2020)

2.8 Greater Sydney and Regional Centres Planning

Greater Sydney embraces 38 separate municipal councils (there are 132 councils over the whole of the state of New South Wales (NSW)). Within Greater Sydney there are over 650 suburbs covering 10,000 sq. km.). The City of Sydney is one of these municipal councils and itself has 33 suburbs. Other urbanising regions bordering Greater Sydney, to the north (Central Coast Region and Newcastle and Hunter Region) and to the south (Wollongong and Illawarra Region). The suburbs of Greater Sydney and Regions are linked by transport and urban infrastructure planned by the State of NSW. Part of this planning is done under the aegis of the Greater Sydney Commission (GSC) (herein called the Commission) (established in 2016) (to be addressed).

Greater Sydney (population at 5 m in 2020) planning is currently under *A Metropolis of Three Cities – The Greater Sydney Region Plan* (2018a). The Plan is outlined via the Greater Sydney Commission at https://www.greater.sydney/structure-plan-metropolis-of-three-cities

The Plan is a 38-year vision (2018–2056) entitled Towards Greater Sydney 2056 (2018b) (herein called Sydney 2056). This plan generally aims to change the structure of the city in accommodating renewal and new growth. The plan was unveiled by the overarching planning body, the GSC and then adopted by the New South Wales (NSW) government. The plan (Plate 2.4) is built around a new concept of three cities (major centres in brackets) within Greater Sydney, being: (1) Eastern Harbour City (currently Sydney City, North Shore beaches and Sydney airport district); (2) Central Rivers City (based around Parramatta, Blacktown and Olympic Park); and, (3) Western Parklands City (Penrith, proposed Western Sydney airport, Liverpool, Camden and Campbelltown). Within these designated 'three cities' are numerous small cities (i.e. City of Sydney was noted above) governed by local councils.

The infrastructure to serve the three cities in concept form is shown in Plate 2.5.

The Greater Sydney Commission states that, under the Three Cities blueprint, each city has a different economic focus. Looking at this designation, the Eastern Harbour City (right) is centred on the traditional CBD, as the centre of financial and

Plate 2.4 Greater Sydney three cities concept. (Source Greater Sydney Commission 2018)

professional services. The blueprint nominates the Central Rivers City around Parramatta (centre in red) a hub of administrative and business services. Within this hub is the Westmead health and education precinct featuring medical and education-related services. Also, by 2056, a Western Parklands City (yellow) (including growth centres) will focus on the new Western Sydney Airport at Badgerys Creek (development commenced in 2019 with an expected airport opening of 2023). The centre closest to the airport, Liverpool, is seen as a future base for trade, logistics and advanced manufacturing industries.

Commenting on Greater Sydney development, the State of Australian Cities 2014-2015 (Australian Federal Government 2016) comments that cities throughout Australia have followed sprawl and cheap land. The report notes, however, this trend is imposing economic and social costs on the country. It's further noted that soaring house prices mean poorer people have to settle for cheaper dwellings on the periphery of the city. The report states that this means the districts' populations face inadequate public transport and long petrol-based commutes to get to work. The State has thus adopted a major planning goal in renewal planning for districts based on the '30 Minute City' concept. This is illustrated in Greater Sydney Plan to 2036

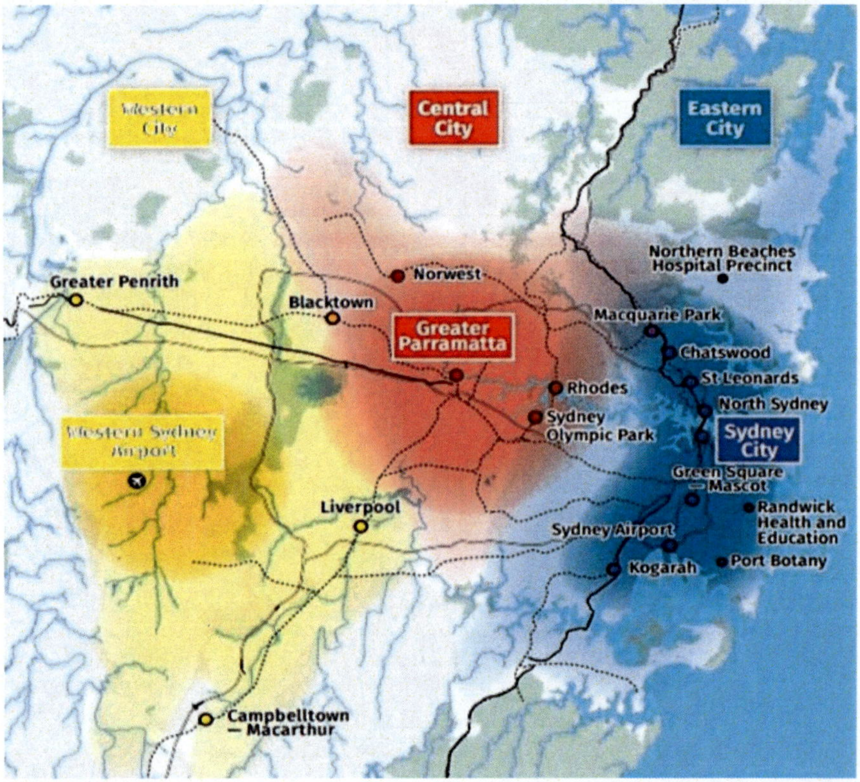

Plate 2.5 Greater Sydney Three Cities Infrastructure Outline (Source Greater Sydney Commission 2018)

(2018d) (herein called the Plan (Plate 2.6) https://www.planning.nsw.gov.au/plans-for-your-area/a-metropolis-of-three-cities/a-metropolis-of-three-cities

The City of Sydney is centered on the larger circle (orange) above (Chap. 3). Note also the Sydenham to Bankstown Corridor in the lower right below Sydney City as a string of eleven rail way hubs (red circles) (Chap. 4).

Value Capture

Moving to value capture, there are different planning models, as reviewed in Chap. 1. The State notes in the Plan that the *Environmental Planning and Assessment Act 1979 No 203* will rebalance growth and deliver its benefits more equally and equitably to residents across Greater Sydney. The State further notes the Plan is the first to be prepared concurrently with Future Transport 2056 and the State Infrastructure Strategy, The State notes the aligning of land use, transport and infrastructure planning in reshaping Greater Sydney (into the three cities concept as reviewed above). Within the Greater Sydney Plan, the Commission also released 20-year plans for the six districts (within three cities) as nominated (Plate 2.7).

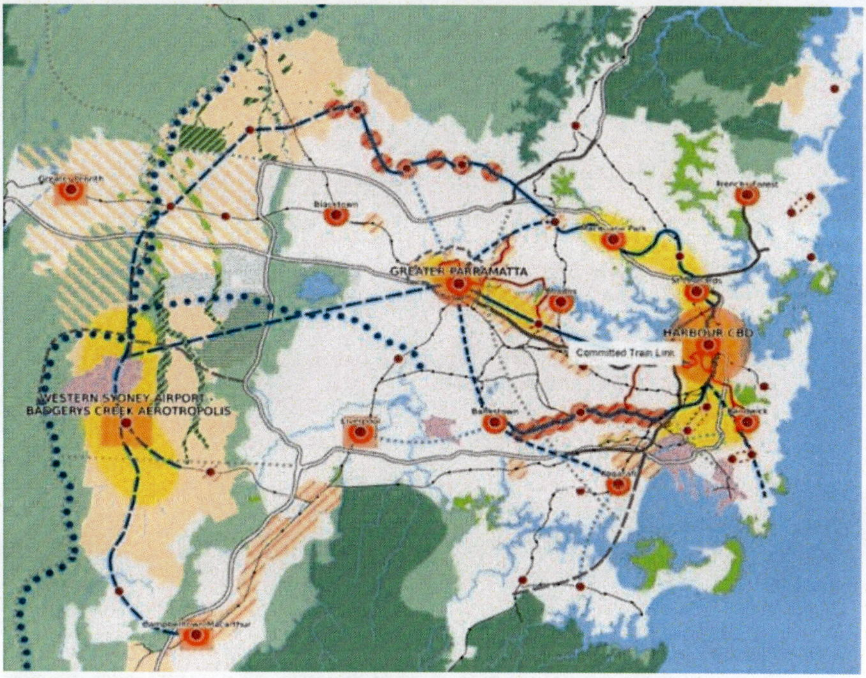

Plate 2.6 Greater Sydney plan to 2036. (Source: Sydney Morning Herald 2014)

The District Plans support the implementation of the Greater Sydney Region Plan. These six districts (each with its own plan) include: (1) Central; (2) North; (3) South; (4) South West; (5) West Central; and, (6) West. The six district draft plans were prepared by the Greater Sydney Commission in consultation with the community, state agencies and local councils. The Central District is addressed (with a case study) in Chap. 3 and the South District in Chap. 4. The Sydney districts' main attributes (as nominated by the State) are as follows (clockwise within Plate 2.7):

Central: The Sydney CBD will continue to be a core for businesses, jobs and financial activity. This district includes the suburbs (a-z) Bayside, Burwood, Canada Bay, Inner West, Randwick, Strathfield, Woollahra, Waverley; and, City of Sydney.

South: This district is also culturally diverse and includes: Canterbury-Bankstown, Georges River, and Sutherland. Priorities include the growth of Kogarah health and education super precinct; and urban renewal within district growth corridors.

South West: This district is one of the most culturally diverse communities in Australia due to its Aboriginal, Colonial and Migrant history. The district includes Camden, Campbelltown, Fairfield, Liverpool and Wollondilly. A key State focus within this district is to plan for a 21st Century aerotropolis surrounding the proposed Western Sydney airport.

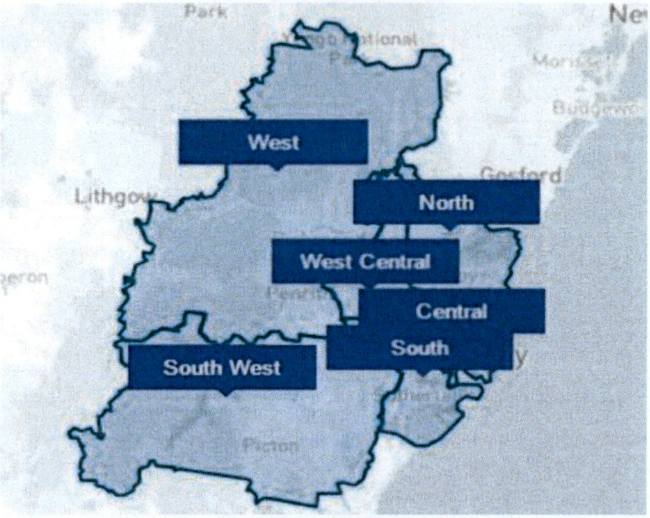

Plate 2.7 Districts within the Greater Sydney Region Plan. (Source Greater Sydney Commission 2018)

West: This district is made up from the Blue Mountains, Hawkesbury and Penrith communities. Priorities include protection of the district's natural landscape, better transport connections and creating a hub for international tourism.

North: This area will include coastline suburbs, bushland neighbourhoods and harbour communities such as; Hornsby, Hunters Hill, Ku-ring-gai, Lane Cove, Mosman, North Sydney, Northern Beaches, Ryde and Willoughby. Priorities include growing jobs in centres and improving public transport.

West Central: This district full of vibrant multicultural communities and has fast growing retail and employment centres. This includes; Blacktown, Cumberland, Parramatta and The Hills. A key priority will be to ensure that Greater Parramatta will fulfil its role as Greater Sydney's Central City.

The State goal was to have coordinated and effective planning in place for land use, transport and infrastructure provisions. All new planning proposals and development applications within a district (above) will need to consider the district plan. Finally, members of local council and the Department of Planning, Industry and Environment (DPIE) have delegated authority to act for the Greater Sydney Commission in certain circumstances for plan making within a district. Noted earlier, the six District Plans will be implemented in conjunction with other plans (i.e. transport and infrastructure) across Greater Sydney. Transport plans (under Transport 2056) (2018c), for example, include projects such as the WestConnex motorway project and the rejuvenation of Parramatta Road. The Commission also states (in 2020) that it is examining issues such as affordable housing (5–10% goal), green spaces and urban renewal. Finally, the district plans will directly inform local council planning and also influence the decisions of state agencies managing the Greater Sydney Region Plan for future growth.

Commenting on Western Parklands City, for example, the Greater Sydney Commission Chief Commissioner (Lucy Turnbull) said at the planning launch (2018) that the new airport (Badgery's Creek) will revolutionise the way people in Western City will live and work. She commented that the Western Sydney Airport will be the anchor for many jobs and underpin the growth of these centres. There is also a State vision that would see most people working within 30 commute minutes of where they live (referred to earlier).

Greater Sydney's 38 municipalities continue to grow and to face urban renewal. It is the State of New South Wales (NSW) planning (via particularly the appointed Greater Sydney Commission) that nominates 'growth and renewal areas' within Greater Sydney. This is done (in theory) in cooperation with respective 38 local government elected bodies within Greater Sydney. Beyond Greater Sydney are regional growth centres as noted earlier. On the border of Greater Sydney these centres include, for example, growth regions of the Central Coast (Chap. 5) and regional capital cities such as Newcastle City (north of Greater Sydney) (Chap. 6) and Wollongong City (south of Greater Sydney). There is and there will continue to be debate on the growth and renewal of Greater Sydney and Regions bordering Greater Sydney. One perspective of future growth of Greater Sydney is offered by the Urban Taskforce Australia (UTA) (Plate 2.8) (noting Greater Sydney's apartment towers).

Moving from Greater Sydney to the Central to Eveleigh Corridor (Chap. 3) the higher densities and introduction of high rise buildings moves from the CBD to immediately south (Plate 2.9). Here the orange in the lower centre represents the State Government plans for renewal of this corridor. The renewal buildings follow the rail line from Central to Eveleigh (bottom), including Redfern and Waterloo.

The transport changes envisaged within this area is contained in Future Transport 2056 (NSW 2018) https://future.transport.nsw.gov.au/ (as noted). This corridor is the central core of the Sydney Metro plans (examined in Chaps. 3 and 4).

On the growth of Greater Sydney, Chris Johnson (CEO of UTA at the time as noted in Chap. 1), was quoted on density and Sydney.

> Key Sydney mayors are backing calls to build hundreds of tall apartment towers near railway stations to avoid the city sprawling towards an urban population disaster. Former NSW state architect Chris Johnson has forecast Sydney needs to build 100 new high-rise apartment towers a year for the next 50 years to protect suburban life from being ruined by overdevelopment as the city's population grows. "Five thousand new towers, containing 110 apartments each, would provide 550,000 new homes — that will accommodate only a third of our planned population growth over the next 50 years," (Daily telegraph 1 Sep 2014)

Johnson continued within the article:

> Many in the community are not comfortable with apartment towers. They feel this threatens the detached house model we have grown comfortable with. But a growing number see apartment living as desirable. There are four major reasons to support such development. First, towers take up a fraction of the land that detached houses do. Second, there is a growing preference for younger people to live in more urban environments with amenities

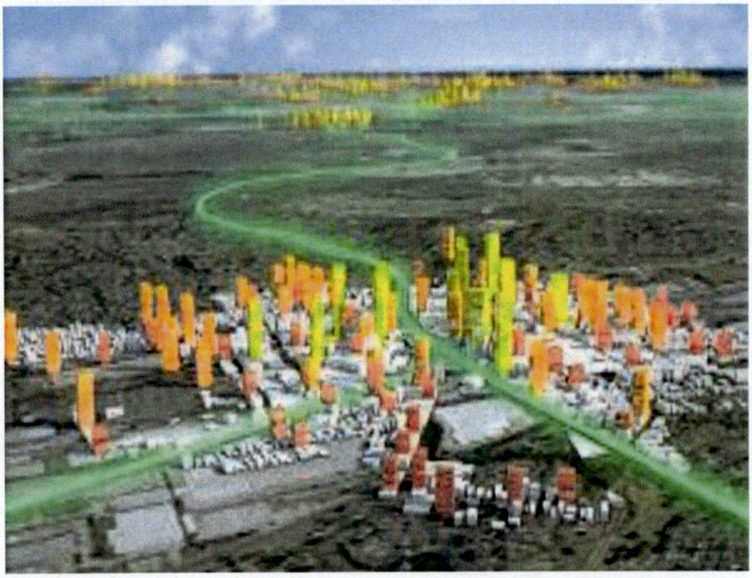

Plate 2.8 Greater Sydney Apartment Towers. (Source Urban Taskforce Australia and Daily Telegraph 1 Sept 2014)

Plate 2.9 Urban renewal strategy central to Eveleigh. (Source NSW State Government 2020)

nearby. Third, the public transport system will work better with more density. Car usage will not be as critical for many people. Fourth, the suburbs will be preserved by focusing development in the centres, and suburban dwellers will have the amenities that come with an urban centre nearby. Clearly, consumers are demonstrating that they want apartment

Plate 2.10 Development Idea for Urban Renewal Area. (Source Daily Telegraph 1 Sept 2014)

> living in towers near transport and we need to make this the norm across the Sydney rail- and light-rail network to house our growing population.

Johnson concluded in citing an artist's concept for renewal under local government guidelines (Plate 2.10). He noted that the development incorporates order and open spaces and mixed residential and commercial development. Many argue that there is too great a reliance on high-rise living (vs medium-rise) for Greater Sydney and Region (Chap. 1).

2.9 Selecting Case Study Areas

Having reviewed Greater Sydney and Regions planning and the Value Capture Planning (VCP) model, four case study areas undergoing renewal or revitalisation in Australia are selected. The research objective is to apply the VCP model across several urban areas to determine results for a wide application of the model. The four study areas selected contain major renewal areas. These include (renewal area and city in bracket): (1) an inner city area (Central to Eveleigh Renewal Area (CERA), City of Sydney); (2) a middle ring city (Sydenham to Bankstown Urban Renewal Corridor (SBURC), City of Canterbury Bankstown); (3) a growth centre city (Gosford City Centre Revitalisation) (GCCR), Central Coast Council); and, (4) a regional capital city (Newcastle City Centre Renewal Area (NCCRA) (City of Newcastle). For each case study renewal area a background of population, planning and development trends are examined. A summary of the renewal or revitalisation case studies by type of urban area, renewal or revitalisation plan, case study precinct and an example relevant planning document is presented in Table 2.5.

Table 2.5 Renewal and revitalisation plans – greater Sydney and regions

Type of urban area	Renewal or revitalisation plan	Case study precinct	Web access (Example of relevant planning document)
1. Inner City	Central to Eveleigh renewal area (CERA)	Waterloo renewal precinct	https://www.planning.nsw.gov.au/Plans-for-your-area/State-Significant-Precincts/Waterloo
2. Middle Ring City	Sydenham to Bankstown urban renewal corridor (SBURC)	Canterbury renewal precinct	https://www.planning.nsw.gov.au/Plans-for-your-area/Priority-Growth-Areas-and-Precincts/Sydenham-to-Bankstown-Urban-Renewal-Corridor
3. Growth Centre	Gosford city Centre revitalisation (GCCR)	Gosford city Centre	https://www.planning.nsw.gov.au/Plans-for-your-area/Regional-Plans/Central-Coast/Gosford-city-centre
4. Reginal capital	Newcastle City renewal area (NCRA)	Newcastle west renewal precinct	https://www.newcastle.nsw.gov.au/getattachment/Development/Land-Use-Planning/Wickham-Master-Plan-Project/Final-draft-Master-Plan-Nov2017_web.pdf.aspx?lang=en-AU

2.10 Surveying Case Study Areas

Having established case study areas, author surveys were then commenced. Surveying case study areas is the first step in determining the adoption of a value capture planning (VCP) approach. The surveys as outlined within the chapters and consist of: (1) visual on ground surveys and photos; (2) reviewing geographical aspects and development trends within and beyond the areas; and, (3) interpreting any renewal or revitalisation plans (noted above) applying to those areas.

2.11 Summary

This chapter outlined the development of a value capture planning (VCP) model and detailed how it would be applied to study areas (4 cities within urban corridors in Australia). Firstly, the potential opportunities for the application of value capture to the renewal of cities was outlined. Attention was drawn especially to the implications of the coronavirus pandemic (2020) on future urban planning (Chap. 1). The VCP model was then outlined and selected planning principles of the model explained. These principles and components (in brackets) were noted as: *Equitable Housing* (affordable, social and market housing); *Public and Open Spaces* (natural areas, open spaces and public spaces); and, *Sustainable Transport* (rail, bus, and active transport). The concept of extending the VCP model application from a precinct level (i.e. renewal areas) to areas beyond the precinct was also commented on.

The Greater Sydney and region centres planning was then introduced. The selected study areas and case study precincts within those areas were outlined. Finally, comment was offered on the surveys to be taken within the case study precincts.

References

Australian Federal Government (2016) State of Australian cities 2014–2015. Federal Government, Canberra

NSW Government (2018) Future Transport 2056. NSW Government, Sydney

Rauscher RC, Momtaz S (2014) Sustainable communities: a framework for planning. Springer, Cham

Part II
Inner City Renewal and Value Capture Planning

Part II addresses inner city renewal and application of value capture planning (VCP) (as outlined in Chap. 1). A VCP model (as developed in Chap. 2) is applied to Greater Sydney's Central to Eveleigh Renewal Area (CERA). The Waterloo Housing Estate Renewal Precinct within CERA (in the Redfern and Waterloo locality) is taken as a case study precinct (Chap. 3).

Chapter 3
Inner City Renewal and Value Capture Planning – Greater Sydney's Central to Eveleigh Renewal Area (CERA)

Abstract The chapter addresses renewal of inner city areas and applies a value capture planning (VCP) model. Two subjects (from earlier chapters) of significance to inner city areas are firstly summarised: (1) the implications of the coronavirus pandemic (2020) for future urban planning (Chap. 1); and, (2) VCP model (including planning principles of: equitable housing, public and open spaces, and sustainable transport) (Chap. 2). The Central to Eveleigh Renewal Area (CERA) (under the State of New South Wales and within the City of Sydney) is then introduced. The VCP model is then applied to the case study Waterloo Housing Estate Renewal Precinct within the CERA. Results and conclusions are then drawn. The application of the model to areas beyond the Precinct is also addressed, including application to: the CERA; the City of Sydney overall; and, other inner city areas of Greater Sydney. At chapter's end it's noted that Chap. 7 will address: (1) overall results of applying the model to each of the four case study areas; and, (2) the application of the model to any urban or regional centre elsewhere in Australia or in the world.

Dedication There are many people in the history of cities and renewal who have left major legacies. This chapter is dedicated to two of them (Plate 3.1), Mum Shirl (social worker and community activist) and Colin James (architect and housing innovator)

Coleen Shirley Perry Smith AM MBE (1921–1998), better known as ***Mum Shirl***, was a prominent Wiradjuri woman, social worker and humanitarian activist committed to justice and welfare of Aboriginal Australians. She was a founding member of the Aboriginal Legal Service, the Aboriginal Medical Service, the Aboriginal Tent Embassy, the Aboriginal Children's Service and the Aboriginal Housing Company in Redfern, a suburb of Sydney. During her lifetime she was recognised as an Australian National Living Treasure. (Source: Wikipedia 2020)

Col James (1936–2013) was a giant who took great strides with gentle steps, followed a career-long path with clear direction and purpose, but always had time for a cup of tea and to listen to those seldom heard by the architectural world he loved. Here was a giant who taught generations of students by using the University of Sydney design studio to work in the community for co-operatives, Aboriginal organisations and the homeless. James completed further study in town and country planning at UNSW under Professor Denis Winston. The course precepts – "serve your client and community equally" and planning based on public good – were obvious in his work. (Source: Wikipedia 2020)

R. Rauscher, *Renewing Cities with Value Capture Planning*,
https://doi.org/10.1007/978-3-030-62958-8_3

Plate 3.1 Dedications: Mum Shirl and Colin James

3.1 Inner City Renewal and Value Capture Planning (VCP)

The City of Sydney (within Greater Sydney) was introduced in Chap. 2. Cities facing social, economic and environmental challenges (especially post COVID19) related to development was summarised. At the same time, cities (including Greater Sydney) spreading ever outwards into growth areas and regional capitals (examined in Chaps. 5 and 6 respectively). The chapter also examined divergences of opinion on city development (commencing in the early 1970's in the City of Sydney in the 'green bans' era as reviewed).

One implication (noted in the Chap. 1) of Covid-19 is a needed review of urban densities and government (and private company) financing of housing, urban infrastructure, public and open spaces, and transport. The value capture planning (VCP) model was also introduced here. It was surmised that new urban development models will become central to renewing cities in the future. For Greater Sydney this means planning renewal for 38 separate municipal councils. The chapter noted that the City of Sydney is one of these municipal councils and itself has 33 suburbs. It is the State of New South Wales (Department of Planning, Industry and Environment) (DPIE) and the Greater Sydney Commission that guide planning. Greater Sydney has a population at 5 m (2020). The current plan is A *Metropolis of Three Cities – The Greater Sydney Region Plan* (2018) as outlined within https://www.greater.sydney/structure-plan-metropolis-of-three-cities. In this planning the City of Sydney falls within the Eastern Harbour City. It was also noted that the designated 'three cities' contain numerous small cities (i.e. City of Sydney) governed by local councils. Greater Sydney and the City of Sydney are looked at more closely next.

3.2 Greater Sydney and City of Sydney

The heart of Greater Sydney is the CBD of the City of Sydney (Plate 3.2).

As noted above the review of planning of Greater Sydney (post coronavirus) is likely to include government and private company financing of housing, public and open spaces, and sustainable transport. The role of value capture planning is thus likely to become central to renewing cities and regions. In considering this, the renewal planning of the City of Sydney's Central to Eveleigh Renewal Area is addressed next.

3.3 Central to Eveleigh Renewal Area (CERA)

The *Central to Eveleigh Renewal Area* (CERA) (Plate 3.3) (first announced by the State in 2014) includes the following renewal area precincts (from top clockwise): (1) Central Station; (2) Lawson St to Cleveland St; (3) Redfern Estate; (4) Waterloo Estate; (5) South Eveleigh; (6) North Eveleigh; and, (7) Redfern Station Precinct. The CERA is within the State's Central to Eveleigh Urban Transformation Strategy (2017) (herein called 'the Strategy'). (infrastructure.nsw.gov.au).

The CERA renewal includes upgrading two rail stations (Central and Redfern) and a proposed new station at Waterloo (construction commenced in 2019 and trains are expected to be operating in 2024). This area remains a significant site within the State's Infrastructure NSW unit https://www.infrastructure.nsw.gov.au/. This unit also acts as the master developer of urban transformation programs, including land

Plate 3.2 Sydney's CBD from Circular Quay looking south to Central and Redfern-Waterloo. (Source: Flickr Sky in the Limit 2014)

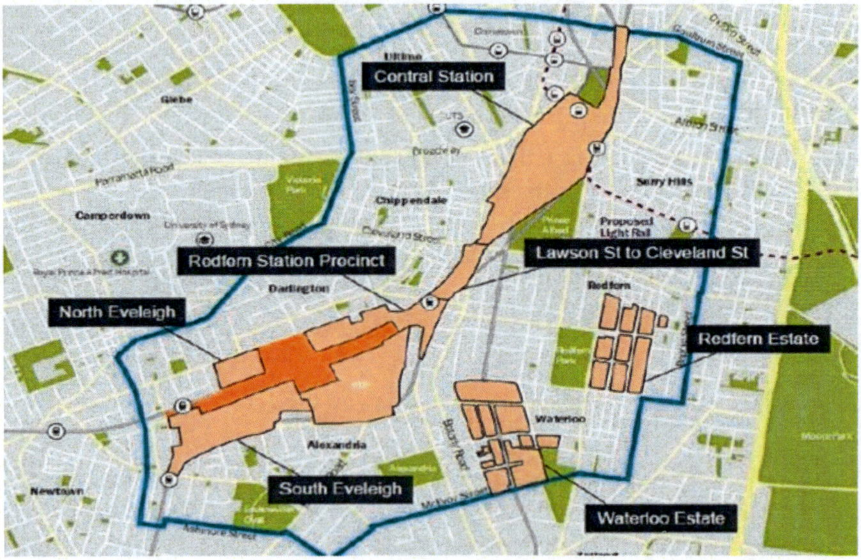

Plate 3.3 Central to Eveleigh Renewal Area (CERA) Precincts. (Source: Infrastructure NSW 2020)

development and transport infrastructure. The State notes that the renewal area precincts (referred to above) consolidates several years of work with government partners to model and analyse the CERA (i.e. such as introducing greater density and diversity of housing). The State also notes that the planned Sydney Metro Rail will travel through the Sydney CBD. The first trains started operating Rouse Hill to Chatswood in 2019. Metro will then have stops at Waterloo, Sydenham and several stops to Bankstown (to be completed in 2024) (see Chap. 4 on this project). Finally, the development at Waterloo and Central Stations will likely affect Redfern Station (given that station's close proximity to the future Waterloo and Central Metro stations).

The State notes that the 3 km long corridor from Central to Eveleigh would open more than one million square metres of development space to the market. This makes the corridor the biggest urban renewal opportunity in Sydney. It was also noted that this scale of project would dwarf the $6 billion Barrangaroo precinct (nearing completion about 2021). It is understood that the NSW Government opened expressions of interest to global companies with expertise in building over train stations in an effort to transform the train tracks into high-rise office, retail and apartments. The CERA strategy also sets a framework for detailed master planning in each precinct (thus with VC potential), especially government owned land sites. The State notes that precinct specific Development Control Plans (DCPs) will set the detailed planning and design guidelines. These will be developed with stakeholders including: City of Sydney; Department of Planning, Industry and Environment (DPIE); and, the local communities.

The State also notes that renewal plans would respond to design excellence and innovative sustainable technology including accommodating: (1) the direction for future growth; (2) changes to land use and built form; (3) public realm, open spaces, community facilities, transport networks and sustainability interventions; (4) the short, medium and long term population, housing and employment projections; and, (5) a clear implementation framework, including funding arrangements (of interest here for the VCP model addressed later in the chapter). The plans will also include designing for: mixed businesses; diverse communities; heritage buildings; cultural opportunities; short distances between work and home; services and recreation; and, education and learning opportunities. Moving from precincts to the whole of CERA a 'massing strategy' (in transformation project zones) defines potential building sizes within the project zones (Plate 3.4) (see also 2.8 in Chap. 2).

There are a number of issues around the CERA planning, including densities. State views on the subject can be found at: http://www.urbangrowth.nsw.gov.au/videos/. On this subject, the State has also worked with the Committee for Sydney (community based group). The group's comments are on its web at http://www.sydney.org.au/?portfolio=density-done-better. Infrastructure NSW has stated that there are obvious trade-offs with higher density neighbourhoods (i.e. many people will feel there is an imposition of taller buildings). The way the State is approaching precincts the State will set the benchmark for how different buildings and structures, old and new, are integrated and used in a variety of ways. The State notes that tall buildings are not appropriate everywhere, but well-designed towers can have a place close to transport and employment centres.

Examining the proposed densities within CERA zones, the Mayor of the City of Sydney, however, notes that Infrastructure NSW plans for Central to Eveleigh involve doubling the population along the rail line from 52,000 to as much as 108,000. The City Council states that no area of the city could sustain the scale of density being planned under this massing strategy (particularly with its large social housing community (i.e. Waterloo and Redfern)). The City Council gives further comment below.

> The State proposed apartment density under CERA at Waterloo is of a scale seen only in pockets of New York or Hong Kong. It's a greater density than anything in Singapore. The

Plate 3.4 Sydney Central to Eveleigh Massing Strategy. *Left to right:* Central to Redfern and Eveleigh; and, Central to Redfern. (Source: Infrastructure NSW 2020)

government is planning a precinct of 20 and 30-storey towers that would increase the density around Waterloo to the equivalent of 70,000 people per square kilometre. That would be more than four times as dense as the current densest area in Australia, Pyrmont, which houses 14,000 people per square kilometre. In comparison, the Green Square area is planned to house 22,000 people per square kilometre. (Source: Council chief executive officer, Monica Barone 2017)

The State however (via the Urban Growth unit) concluded Waterloo would be less dense per hectare than developments at Green Square, Central Park and Darling Square. Moving to environmental sustainability, some of the key CERA guidelines the State has adopted are in Addendum 3.2. The key precincts within CERA examined next.

3.3.1 *Central Precinct*

Looking at the northern part of CERA (top of plate), the government has broken down the 24-hectare Central Precinct, into a series of sub-precincts. For example, Sydney's rail yards (Plate 3.5) between Central Station and Cleveland Street, is set for renewal with office towers, parks, shops and potentially residential buildings (being planned by the State). At the western gateway of this precinct, the government says the area has the potential to kick start precinct development. An example of the proposed transformation of this precinct is illustrated in the plate. The artist impressions show (left to right) the proposed new Metro Sydney section of Central Station (left) and the potential of high rise buildings over the rail (right).

The new Sydney Metro is a major State investment project as illustrated in Plate 3.6. Here is shown (left to right): an artist impressions of the proposed new Sydney Metro section of the Central Station Precinct (left); and, the potential of constructing high rise commercial buildings within the Precinct (right).

Plate 3.5 Central Station Precinct Transformation Proposal. *Left to right.* Current rail station and yards; and, artist vision of high rise buildings cluster. (Source: Infrastructure NSW 2020)

Given the importance of this Central Station Precinct, the Lord Mayor of the City of Sydney, Clover Moore, welcomed the above Precinct plans noting:

> Forward thinking development is needed in this under-utilised part of the city. For years, debate has raged over what we should do with the land above and around the Central Station rail yards. Now NSW Planning and Atlassian (proponent) have a plan that the Council is excited about. This would be a high tech precinct above the rail yards, connecting the surrounding suburbs with new walking connections. (Source: Sydney City Council Media 2019)

The State in responding also noted in respect to this Precinct that:

> Australia's largest tech company (Atlassian as noted) will anchor a 35-storey skyscraper above a new Valley-esque technology hub next to Sydney's Central Station. More than a year after premier Gladys Berejiklian first flagged plans to transform the southern end of the CBD into a tech and innovation hub, the NSW government released the first images of its vision for the 24-hectare Central Precinct. The proposal sets out the urban renewal of the precinct into a transport-centered commercial district. This includes, improving public connections to bordering suburbs Chippendale, Surry Hills, Ultimo and Redfern and providing green walkways to Belmore and Prince Alfred parks. The southern central Sydney location provides a natural extension to the city's development. The opportunity is comparable [to] large urban renewal programs including Kings Cross, Euston and Olympic Park in London and Hudson Yards in New York. (Source: Infrastructure NSW Media 2019)

Moving from Central to the planned Waterloo Station, Urban Developer (Sydney Metro) notes the State secured approval for the proposed development above the station. A number of building options were investigated for the project, including initiatives in Brisbane https://theurbandeveloper.com/articles/ferny-grove-station-redevelopment The Waterloo proposal would include four residential apartment buildings ranging from 23 to 29 storeys. The development would also comprise mixed uses such as retail and commercial office space (addressed further under 3.5).

Plate 3.6 Artist impression of proposed new Metro Sydney at Central and high rise buildings. *Left to right:* proposed new Metro Sydney Central Station; and, the potential high rise buildings. (Source: Infrastructure NSW 2020)

3.3.2 North Eveleigh Precinct

Looking at the North Eveleigh Precinct within CERA, the State comments as follows.

> North Eveleigh (Plate 3.7) will be the first precinct developed as part of transformation efforts on government-owned land. People in the mainly residential precinct of North Eveleigh could enjoy easy, close access to cultural, education and work-related activities. The design includes repurposed heritage buildings, creating a distinct character that draws upon its industrial heritage and provides space for local shops and community facilities. The precinct will be a sociable, with a new park and features that build on the strong local arts, culture and history of the site. Taller buildings will be adjacent to the rail corridor with lower buildings respecting the character of the surrounding area. (Source: Infrastructure NSW Media 2019)

3.3.3 South Eveleigh Precinct

Moving to the South Eveleigh Renewal Precinct the Mirvac-lead consortium has been developing (commenced 2016) within the precinct (i.e. within the Australian Technology Park (ATP). This included a $1 billion office and emerging technology precinct (close to the heart of Sydney's CBD). Plate 3.8 shows the open spaces portion of the planned development at South Eveleigh.

The South Eveleigh project also included construction of a $2.1bn Tech Incubation Fund built in conjunction with the Commonwealth Bank and Centuria. The Minister at the time indicated that the consortium had made a commitment to improving community facilities within the ATP. This included an investment to

Plate 3.7 Artists' Impression of Developments at North Eveleigh. *Left to right:* High rise with heritage building in foreground; and, high rise with medium density in foreground. (Source: Infrastructure NSW 2020)

Plate 3.8 South Eveleigh Artist Impressions. *Left to right:* green pedestrian areas; and, open space plaza. (Source: Infrastructure NSW 2020)

upgrade public areas, making them more accessible and including a community building, childcare facilities, fitness facilities and the provision for retail outlets, cafes and restaurants. In accordance with the government's requirements, it is understood that the consortium will also maintain and manage the site's heritage buildings, and ensure continued and safe public access to important destinations like Redfern railway station. Proceeds from the State sale of ATP was to be invested in the local area, helping finance major infrastructure investments in the Central to Eveleigh corridor.

Moving beyond the ATP and still within to the South Eveleigh Precinct, the State comments that the renewal transformation could create a diversity of housing in South Eveleigh. Potential new housing includes renewed social housing and additional affordable housing. The State further notes that there could be a mix of apartment buildings, including taller residential buildings alongside the rail corridor and lower buildings on the precinct's edges in transition to the existing neighbourhood (also noted within North Eveleigh above). A key question in planning this precinct is where the funding of social and affordable housing and facilities under urban renewal will come from, as examined next?

3.3.4 Funding Urban Renewal

Moving to funding of CERA, the State notes that the costs of community facilities will be borne by: local councils, the State, land owners and developers. The State notes that it is investigating a broad-based approach to development levies in areas of increased development intensity. Expanding on this the State notes there is the potential to apply a contributions system to help fund new infrastructure, but also retain the commercial viability of projects (Chap. 1). The State also intends to use the CERA Strategy to help guide development contribution planning during the precinct planning stage. The State notes that given the movement of people, significant government investment in the road and pedestrian network will be required. In addition the State notes the existing Affordable Housing Levy (i.e. Green Square, Sydney) that provides a funding source for affordable housing could be strengthened and applied to CERA. The State concedes that the proceeds from land sales or proceeds from the sale of development rights can also be used to fund new social housing programs (see Chaps. 1 and 2). Finally, the State notes that a value capture mechanism will be introduced which incorporates a contribution of $20,000 for each new apartment. This would help fund the train station and affordable and social housing. This approach is similar to the Parramatta Light Rail project. It is understood that the Parramatta contribution will be forwarded onto the private investors of the light rail project. With the likely rise, however, in housing prices and rents there is a risk that families will not be able to afford the Parramatta (or Redfern and Waterloo locality) area and be pushed to the outer suburbs. Having looked at several precincts and funding aspects, the Redfern and Waterloo locality is introduced next (before the case of Waterloo Precinct is examined).

3.4 Redfern and Waterloo Locality

The Redfern and Waterloo locality is a highly significant urban place in the City of Sydney. A review of this locality follows, including: background; residential renewal; and, transport.

3.4.1 Background to Redfern and Waterloo

An examination of the geography of Redfern and Waterloo (Plate 3.9) is a starting point to understand the locality. Shown here is the outline (left frame) of Central Redfern with Waterloo; the high density of the locality (centre frame); and, the large

Plate 3.9 Redfern and Waterloo Locality. *Left to right:* Map indicating Central, Redfern with Waterloo; aerial Illustrating density of locality in foreground; and, Redfern and Waterloo public housing in orange. (Source: Infrastructure NSW 2020)

amount of public housing (right frame) with Waterloo (bottom left) and Redfern (right top) housing (orange).

Examining local history, the Aboriginal Gadigal people were the first inhabitants of Redfern and Waterloo locality dating thousands of years ago (known as Gadigal Country). To appreciate this aspect, a timeline of some of the activities of the Aboriginal community at this location is summarized in Addendum 3.1. To follow the progression of the Redfern Aboriginal Housing Project (planning commenced in 1974) gives an insight into the importance Redfern is to the Aboriginal community (Plate 3.10).

Properties owned by the Aboriginal Housing Company were demolished in 2004 to make way for the Pemulwuy Project (work commenced in 2019), a development offering affordable housing for 62 families, a gymnasium, commercial and retail space, a gallery, accommodation for several hundred students and a childcare space for 60 children. A 99 year lease for the land has been arranged with the Aboriginal Housing Company for over 300 housing units (a percentage to be allocated to students with aboriginal backgrounds).

The controversy (delays getting the housing built) led in time (2018) to the establishment of an Aboriginal Embassy. The embassy drew attention to delays in the project and the final plans not honouring the original intent and quantity of that housing. A member of the 'embassy, Jenny Monroe (left frame) stands beside the Aboriginal flag in 2018 (making a point of the need for housing action). In time (top middle) construction started (late 2019) with a high rise student housing complex (background) and the medium density Aboriginal housing estate (foreground). The student housing is on a 100 year lease (top right) arrangement with the Redfern Housing Company. An artist sketch of the project community centre

Plate 3.10 Redfern and Aboriginal Community. (Source: Infrastructure NSW 2020)

and multipurpose building is shown in the bottom left frame. The streetscape showing the existing terraces and the student housing is shown (bottom centre). A concept distant view illustrates student housing (bottom right). More information on the inner city Aboriginal community can also be accessed at: http://www.sbs.com.au/nitv/the-point-with-stan-grant/article/2016/03/09/inner-sydneys-aboriginal-community-fear-they-are-being-pushed-out-white-hipsters

Moving to 2020s, within the Waterloo and Redfern locality there are the precincts of: (1) Waterloo Housing Estate; and, (2) Metro Sydney South West – Waterloo Station (under CERA) (Table 3.1a). These are State owned lands. Within this precinct (west and adjacent to the housing estate) four residential towers are planned (examined later in Chapter).

3.4.2 *Residential Renewal*

The Redfern and Waterloo locality residential renewal and Waterloo Metro Station development are a central part of CERA. The number of dwellings in Redfern under the CERA, for example, could increase by an additional 22,000 by 2036 (State government residential forecasts). This addition of dwellings to the housing stock is a major driver of population growth in the area. The State says this would provide opportunities for households to relocate from other areas and for new households to form locally. Cost of living in Redfern, however, today (2018) is usually more than an average family's budget. A home can cost $1.5m and rents can exceed $710 per week for a two bedroom place. Here are some examples of housing and rental costs and availability of housing at Redfern in 2017 (Table 3.1b).

It's interesting to note the buyer visits per property at Redfern in May 2017 represented a high demand market at 1116 visits per property (vs. average of 774 per property for the State). Examining Redfern, the Sydney Morning Herald notes the stages of gentrification and the Aboriginal Housing company (as noted above) at: http://www.smh.com.au/nsw/clover-moore-alarmed-by-waterloo-apartment-plans-that-dwarf-singapore-20160516-gowfr0.html

The Sydney Morning comments on gentrification in Redfern as follows:

> Redfern is in the later stages of gentrification and is in a perfectly positioned city location. Conveniently located close to Sydney's CBD (it's just one stop away from Central Station by train) together with a well-serviced bus network. Just 3 km to Sydney CBD, Redfern is

Table 3.1a Redfern and Waterloo Locality Precincts under the Central to Eveleigh Renewal Area (CERA)

Renewal plan	Aspect of plan (example)	Web access
Waterloo Housing Estate	Public Housing Estate Renewal	https://www.planning.nsw.gov.au/Plans-for-your-area/State-Significant-Precincts/Waterloo
Metro Sydney South West Waterloo	New Waterloo Rail Station	https://www.sydneymetro.info/station/waterloo-station

Table 3.1b Redfern housing costs and rentals May 2017

Housing costs $Au		Rental costs $Au/pw	
2 Bed Room	$1.53 m	2 Bed Room	$710
3 Bed Room	$1.61 m	3 Bed Room	$1000
4 Bed Room	$2.27 m	4 Bed Room	$1300

Source: SMH 17 May 2017

Plate 3.11 High rise developments at Redfern. (Source: Infrastructure NSW and Ray Rauscher 2020)

> a suburb on the rise. It's an urban hotspot that's transforming into a favoured destination for young professionals, downsizers, artists and urbanites. Eveleigh Farmers Markets, the Carriageworks Arts Centre as well as Sydney University, UTS and the Australian Technology Park are all within walking distance. Getting around the city is effortless with excellent bus links at the doorstep as well as an ever-growing network of new cycle ways that connect Redfern to Surry Hills, Alexandria, Waterloo and Green Square and the CBD. Over the past decade, the inner-city suburb has morphed into a vibrant urban village celebrated for its sense of community and creative vibe. This inner-city pocket is a thriving creative hotbed reminiscent of New York's Upper West Side, with galleries and studio spaces defining the urban landscape. Warehouse and factory buildings are reborn as studios, design stores and workshops for the ever-growing arts scene, and two significant cultural hubs. (Source: SMH 17 May 2017)

Within the 'urban village', examples of new high rise residential buildings abound at Redfern, (clustered around the rail station) and Waterloo. Recent high rise developments in Redfern are shown in Plate 3.11.

Likewise high rise residential buildings have been built or proposed over the last couple of decades (to 2020) at Waterloo, especially along Botany Rd (examples in Plate 3.12).

Further references to housing and real estate in the Redfern and Waterloo locality can be further assessed at the following web sites (random selection and summary of each in column 3) (Table 3.2).

A demand from the expanded student housing market has resulted in the Iglu Redfern project (#6 in above table) commenting:

> This is a new 18 storey student accommodation building in Redfern's town centre. The Bates Smart-designed building will be Iglu's third facility in Sydney. It will house 370 students close to the major CBD education campuses. The project responds to the growing demand for purpose-built student accommodation in Sydney. A recent study prepared by

Plate 3.12 High rise developments proposed or built at Waterloo. (Source: Infrastructure NSW 2020)

> Jones Lang LaSalle, Australian Student Accommodation 2015 Market Update, finds that there is a shortfall of approximately 29,000 student accommodation beds in Sydney beyond the current development pipeline. The development will allow the restoration and adaptive reuse of the historic Regent Street retail terraces on the site, and create the potential for future laneway activation. (Web Iglu Project 2020)

3.4.3 Transport

Looking at planning transport, major changes are planned for both Redfern and Waterloo. An upgrading of Redfern rail station (Plate 3.13) was proposed by the State in 2017 (entry improvements done in 2019). The left frame shows the position of the station (pink) relative to the surrounding buildings. An artist sketch of the exterior illustrates the proposed upgrade (right). In the process, important factors guiding the strategy (the State noted) included: social and economic development (particularly property development); the sociology of the area; and, the station's relationship to the economic activity in the area.

3.5 Waterloo Renewal Precincts

The Waterloo Renewal Precincts will be next examined under the following parts: Waterloo Metro Quarter Development Precinct; and, the Waterloo Housing Estate Renewal Precinct. In addition two related subjects are addressed: State Master Planning of Waterloo Housing Estate; and, Sydney City Council's Waterloo Estate Renewal Plan.

Table 3.2 Sample real estate market sources 2019 (a-z) in Redfern and Waterloo Locality

<table>
<tr><th>#</th><th>Real estate source (a-z)</th><th>Summary</th><th>Website</th></tr>
<tr><td>1</td><td>Domain (under construction in 2019)</td><td>Accommodating high rise student housing within The Block, Redfern</td><td>https://www.domain.com.au/news/redferns-star-keeps-rising-as-residents-grapple-with-development-20150528-gh9nd6/</td></tr>
<tr><td>2</td><td>Forecasting Study – Sydney</td><td>Projections on Sydney residential development</td><td>http://forecast.id.com.au/sydney/residential-development?WebID=190</td></tr>
<tr><td>3</td><td>General Real Estate – Redfern</td><td>Real estate and neighbourhood of Redfern</td><td>https://www.realestate.com.au/neighbourhoods/redfern-2016-nsw?gclid=CNvv_trogtQCFdcRvQodqfwJhw&rsf=google&rsfdetail=sttF1Ox9u_dc|pcrid|76141259086|kword|redfern|match|p|plid|&s_kwcid=AL!4401!3!76141259086!p!!g!!redfern&ef_id=WMbxDwAABSvFigZp:20170522055721:s</td></tr>
<tr><td>4</td><td>Office Towers Conversion, Redfern</td><td>Office tower design</td><td>http://www.planning.nsw.gov.au/News/2016/Redfern-office-towers-to-be-redesigned-into-residential-apartments</td></tr>
<tr><td>5</td><td>Property Chat – Redfern</td><td>Major social and property changes in Redfern</td><td>https://www.propertychat.com.au/community/threads/redfern-major-changes-coming.11040/</td></tr>
<tr><td>6</td><td>Student Housing Iglu – Redfern</td><td>New high rise student housing with commercial below at Redfern Rail Station</td><td>https://iglu.com.au/iglu-redfern-phase-2/</td></tr>
</table>

Sources: Varied as Noted Above

Plate 3.13 Redfern Railway Station Planning. *Left to right*: Artist sketch of exterior of proposed upgrade; and, model showing the position of the station within central Redfern. (Source: Sydney City Council 2017 and Infrastructure NSW 2020)

3.5.1 Waterloo Metro Quarter Development Precinct

The Waterloo Renewal Precincts examination commences with the Waterloo Metro Quarter Development Precinct (separate development to Waterloo Housing Estate Precinct as examined next). The completion (2024) of the Sydney Metro line (Plate 3.14) (Chatswood to Bankstown) would accommodate a Waterloo Station.

Waterloo Metro Quarter (Plate 3.15) is the area above the future Waterloo metro station (upper left) (between Cope, Botany, Raglan and Wellington Streets).

The proposed Waterloo rail station entry and proposed residential towers are noted in Plate 3.16. Note the spaciousness of the plaza entry and the height of the four high rise residential towers in background.

Finally, the proposed Waterloo station and public spaces are illustrated in Plate 3.17. From left to right shows: artist's impression of activity around the station; a sketch of entry off Botany Rd.; and, Botany Rd. view of high rise buildings and restored church to right.

The State in respect to the Waterloo Station project noted as follows:

> The Government of New South Wales (NSW) has awarded a contract worth $299m to a joint venture (JV) of John Holland and Mirvac for the construction of new Waterloo metro station and an integrated station development. Under the contract, John Holland will be responsible for the delivery of Waterloo metro station while Mirvac will be responsible for the integrated station development component delivery. (NSW Government Media 2019)

Mirvac indicated at the time the plan includes residential, office and retail space. At least 5% of houses will be affordable homes and 70 apartments will be reserved for social housing. It also includes new community facilities, more trees planted and two new public plazas at Cope Street and Raglan Street. UrbanGrowth NSW was given (2018) the green light for its over-station development project, the Waterloo Metro Quarter. The 70,000 sq m precinct, set to be built atop the new metro train

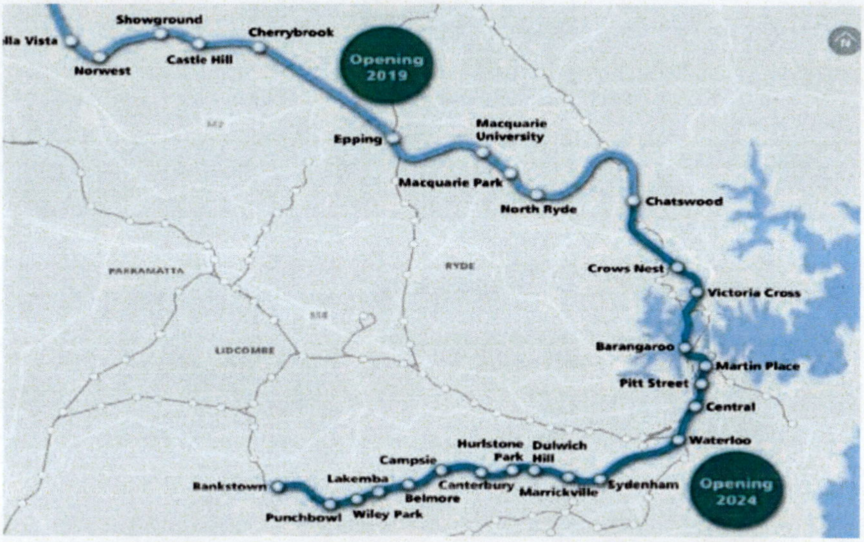

Plate 3.14 Sydney New Metro Rail Line. (Source: Sydney Morning Herald 2014)

Plate 3.15 Map showing Future Metro Sydney Waterloo Station, Metro Quarter and Waterloo Estate. (Source: Infrastructure NSW 2017)

station in Waterloo will yield 700 apartments across four residential towers (as noted). The development will also comprise about 4000 sq m of retail and 9000 sq m of commercial space, shared offices and health services. It is part of the Waterloo State Significant Precinct, a 20-hectare area. Waterloo is one of several city Metro sites, including Barrangaroo, Martin Place and Pitt Street in Sydney's CBD, and north of the CBD at North Sydney and Crow's Nest, where large mixed-use developments will occur above metro stations (Chatswood to Bankstown). The government intends for the Metro Quarter residential component at Waterloo to be

Plate 3.16 Photomontage sketch of Waterloo Urban Project station entry. (Source: SMH 2014)

Plate 3.17 Proposed Waterloo Railway Station and Public Spaces. *Left to right:* artist's impression of night activity around the station; sketch of entry off Botany Rd.; and, Botany Rd. view of entry and restored church to right. (Source: Infrastructure NSW 2020)

completed by 2025. A final point to note, the NSW Government owns the Metro Quarter site and Infrastructure NSW is developing the project (Plate 3.18).

Infrastructure NSW and Sydney Metro have prepared a development masterplan for the Waterloo Metro Quarter. In addition, the State has prepared a state significant precinct study (http://planspolicies.planning.nsw.gov.au/index.pl?action=view_job&job_id=9732). Details included: (1) transfer the City's planning powers for Waterloo metro quarter to the Minister for Planning; (2) change the planning controls to allow building heights of up to 116.9 m (29 storeys) instead of the current 15 m (4 storeys); (3) allow a floor space ratio of 6.1:1 instead of the current 1.5:1; (4) develop the site for mostly residential apartments, with some commercial, retail and community uses; and, (5) provide a plaza to connect the Waterloo metro quarter to Cope Street.

Sydney Metro prepared a development application for the Waterloo Metro Quarter http://www.majorprojects.planning.nsw.gov.au/index.pl?action=view_job&job_id=9393. This was determined by the Minister for Planning. The

Plate 3.18 Concept proposed for Waterloo Metro Quarter. (Source: Infrastructure NSW 2020)

development application was dependent on the Minister approving the new planning controls proposed by Infrastructure NSW in the State Significant Precinct Study. Having been introduced to the Redfern and Waterloo locality and Sydney Metro Waterloo, the Waterloo Housing Estate Renewal Precinct case study is introduced next.

3.5.2 Waterloo Housing Estate Renewal Precinct

Moving from the Waterloo Metro Quarter, the Waterloo Housing Estate Renewal Precinct is a project to span 20 years (Plate 3.19).

Once the planning processes are completed the redevelopment area will be spilt up into super blocks. These are likely to be put to tender for the private sector to redevelop the individual blocks. The proposal would involve the demolition of the existing Waterloo Housing Estate. The Estate dates back to 1970s and incorporates: 2 × 30 storey towers; 4 × 16-storey towers (two of which are shown on the left in the above plate); and, 3 storey walk ups (to the right in the above plate). This totals 2000 current housing units. The existing accommodation is considered by the State as outdated (i.e. maintenance issues).

The suburb of Waterloo is thus undergoing one of the largest urban renewal projects in Australia. The entire development area is approximately 40 hectares with 13

Plate 3.19 Part of Waterloo Estate from the air. (Source: Infrastructure NSW 2017)

hectares owned by the Government. The Council was proposing (as noted by council) an alternative approach to the NSW Government's redevelopment of the Waterloo housing estate. The estate covers an area bounded by Phillip, McEvoy, Pitt and Cope streets. The state government announced that the proposal would see the number of homes triple to 7200 homes, two-thirds of which will be in 17 towers up to 40 storeys high.

The project seeks to create a mixed-housing precinct which will accommodate an additional 30,000 residents and 10,000 dwellings with 1/3 being social and affordable housing while the remainder being private. Land and Housing Corporation (LAHC) (part of NSW Department of Family and Community Services) has been consulting with the community as the master plan options for Waterloo Housing Estate are developed. Three masterplan options were published by LAHC (see community input on development options at Communities Plus website).

Examining further, LAHC published a Visioning Report (May, 2018) which outlines the community's feedback to date. This report is available at https://www.communitiesplus.com.au/major-sites/waterloo The State Significant Precinct Study summarizes the project and includes a masterplan. The existing housing tenants will have the opportunity to return to the public housing once it is constructed. The State notes that one of the main challenges facing the Waterloo Estate Renewal Plan will be the temporary relocation of the existing public housing tenants during the redevelopment. A large portion of the tenants have been at the Waterloo Housing Estate long term and are attached to the locality. The tenants may have fears of being driven out of the area by gentrification. The State expects the development will create a safer environment (an incentive for these tenants to return).

The State also notes that the metro station (addressed earlier) will create the opportunity to transform Waterloo into a better place to live for future and existing residents. It's also noted that although Waterloo is only 4 km from the city, it is relatively isolated with regard to public transport services. The station is also expected to bring jobs to the area while it will provide direct links to key employment areas around the new Metro rail stations at Barrangaroo, Marin Place, Pitt St and Central as noted (and beyond to the planned upgrade of Sydenham Metro station) (see Chap. 4).

3.5.3 State Master Planning of Waterloo Housing Estate

The State's (NSW) Land and Housing Corporation has released its preferred masterplan for the redevelopment of the Waterloo housing estate (https://www.communitiesplus.com.au/major-sites/waterloo). The plan was evolved by the Dept. of Planning Industry and Environment (DPIE) in cooperation with Transport NSW and the Office of Government Architect and the City Council (as a member of the project review panel). The approach taken by the State in engaging the community is contained in Addendum 3.3.

The State notes further:

> Sydney's population is expected to double over the next 40 years. There is thus a need to consider how best to use inner-city land to not only create housing but renew these areas to make them great, liveable neighbourhoods. The State wants to design the renewal to integrate housing with the community. Under the plans for the site, the Waterloo Estate would be transformed into a high-density housing precinct. The 2,012 social housing homes in the estate would be replaced by up to 7,200 dwellings, built over 20 years, making it one of the highest density housing precincts in the country.

Finally, the State plans for Waterloo renewal calls for a number of high rise buildings, community facilities and open spaces (Plate 3.20). The Sydney City Council alternative plan to the State masterplan is examined next.

3.5.4 Sydney City Council's Waterloo Housing Estate Renewal Plan

The Sydney City Council put forward an alternative approach to the State redevelopment plan of the Waterloo, available at this site below.

https://www.cityofsydney.nsw.gov.au/vision/changing-urban-precincts/waterloo-precinct

At a public meeting on 6 March 2019, the City's director of planning, development and transport, Graham Jahn gave a presentation on the City's alternative approach to the Waterloo housing estate (https://www.cityofsydney.nsw.gov.au/vision/changing-urban-precincts/waterloo-precinct)

Though the site is public land (the Council pointed out) the government was proposing 65% of the homes will be private housing, with 30% dedicated as social

Plate 3.20 Waterloo renewal plan visual. (Source: Infrastructure NSW 2020)

housing and 5% as affordable. The government's proposal included 2 parks (however the Council argued that these parks would be overshadowed by the proposed towers). The City of Sydney's approach included: (1) up to 5300 homes; (2) retaining the 30 storey Matavai and Turanga towers; (3) buildings varying from 4 to 13 storeys; (4) 2.2 hectare park as the centrepiece; and, (5) 50% social housing, 20% affordable housing, and 30% private housing. A significantly larger park (Plate 3.21) would be the centrepiece of Council's approach for the precinct.

The park would occupy 2.2 hectares and receive more than the minimum required levels of sunshine. The City's alternative approach for the Waterloo housing estate features buildings of 12–13 storeys. The majority of the rest of the buildings would be 7–8 storeys. The Council offered in its plan to the State examples on how open spaces would be created (Plate 3.22). The Council explained in its report that there would be a widened, tree-lined George Street designed to be safe for people walking and riding bikes. It is here that buildings would be located (7–8 storeys) with shops and cafes on the ground floor.

Information web sites to gain further background on Waterloo are outlined in Table 3.3 below.

More information can also be found at: http://www.centraltoeveleigh.com.au/area/waterloo-estate/#sthash.ubdEBI3Q.dpuf A background on Waterloo and its development is outlined as follows by Infrastructure NSW commenting:

> There is an opportunity to demonstrate great public outcomes in this area of Sydney led by new and more social housing on the Waterloo Estate. The new Sydney Metro station here will be the centre-piece of the transformation. The key consideration will be the redevelopment of the Waterloo Estate. This will be planned with the community to retain a diverse and vibrant neighbourhood into the future. The Department of Family and Community Services (FACS) will renew all existing housing and build more social housing to meet people's needs, alongside affordable and private housing. The transformation will be staged gradually over 15 to 20 years. This could lead to around 5,000 additional homes on the site alongside new and more social housing to replace the 2,000 (approx.) existing social housing dwellings (total thus of 7,200 housing units).

Plate 3.21 City of Sydney Council's proposed park in renewal of Waterloo Estate. (Source: City of Sydney 2019)

Plate 3.22 Artist's impressions of open spaces in proposed renewal of Waterloo Estate. (Source: City of Sydney 2019)

The Value Capture Planning (VCP) model application to the Waterloo Housing Estate Precinct is addressed next.

3.6 Value Capture Planning (VCP) Model Application

The Value Capture Planning (VCP) model (Tables 2.1, 2.2, 2.3, and 2.4 in Chap. 2) could be considered by the State and the City of Sydney Council for Waterloo renewal. That model is now outlined and applied to the Waterloo Estate Renewal Precinct within the CERA (reviewed earlier). The three VCP model planning principles as introduced in Chap. 2 are applied, including: *Equitable Housing*, *Public and Open Spaces*, and *Sustainable Transport*. The second stage of the model, Developer Provisions and/or Levy Alternative is the applied. Additional urban

Table 3.3 Background references on renewal of Waterloo (a-z)

#	Source	Summary	Website
1	Redfern Waterloo Authority (Glenn Searle, author)	History of Redfern- Waterloo Authority	http://www.crossart.com.au/images/pdfs/Glen_Searle_RWA.pdf
2	ABC News on public tenants	Around Waterloo Station	http://www.abc.net.au/news/2016-07-03/nsw-public-housing-tenants-cry-foul-over-waterloo-plans/7565660
3	Inner Sydney City Voice (Community Based NGO)	Waterloo redevelopment comments	http://www.innersydneyvoice.org.au/pub/waterloo-redevelopment-storm-brewing/
4	Inner Sydney City Voice Journal	Artists Impression of renewal	http://www.innersydneyvoice.org.au/pub/wp-content/uploads/2016/05/ISV_WINTER_2016_Web.pdf
5	Redwatch Community Group Newsletter	Issues around public housing	http://www.redwatch.org.au/issues/public-housing
6	South Sydney Herald	Waterloo Renewal	http://www.southsydneyherald.com.au/redeveloping-waterloo-here-we-go-again/#.WR6b4GZdDIU
7	News Dot Com	Waterloo and Densities	http://www.news.com.au/finance/real-estate/plans-for-australias-most-densely-populated-suburb-have-the-locals-up-in-arms/news-story/18f99dc0e363fc4f6b54281c5f128f7b
8	Waterloo Transport	New Metro Line	http://www.governmentnews.com.au/2015/12/sydneys-new-metro-line-to-raze-infamous-waterloo-towers/

Sources: Web links as noted

planning components can be added to or modified jointly between the State and local council. Each of the planning components as selected is now examined.

3.6.1 Equitable Housing

The *Equitable Housing* component of the value capture model for the Waterloo Estate Renewal Precinct is illustrated Addendum 3.4 (Table 3.4). In addressing housing types special attention is given to affordable (20%) and social housing (50%). Given the site is government land a mix of market housing (30%) can also be accommodated.

Table 3.4 Value Capture Planning (VCP) Table: Equitable Housing – Waterloo Housing Estate Renewal Precinct

Step	1	2	3	4
Checklist	Housing types	Meet criteria of need for housing types (Examples)	Address standards and design guidelines for housing types (Examples)	Projected # and % of new housing units (Example)
	New housing on 13 ha of State Govt. land	Various	Various	7200 (100%)
✓	1. Affordable housing	Special housing groups	Innovative architecture	1440 (20%)
✓	2. Social housing	Social support needs	Maintenance efficiencies	3600 (50%)
✓	3. Market housing	Housing as income compatible	Range of housing style choice	2160 (30%)
Steps as per columns				
1	Tick the checklist for housing types that apply			
2	Adopt criteria for housing provision			
3	Refer to design guidelines for housing			
4	Projected units of new housing and % by type			

3.6.2 Public and Open Spaces

The *Public and Open Spaces* component (with elements of natural areas, built open spaces, and public areas) of the value capture model for the Waterloo Estate renewal area is illustrated Addendum 3.4 (Table 3.5). Key factors here include projected public and open spaces in categories of: natural areas; built open spaces; and, public areas (quantities to be determined by final plans).

3.6.3 Sustainable Transport

The *Sustainable Transport* component of the value capture model for the Waterloo Estate renewal area is illustrated Addendum 3.4 (Table 3.6). Key elements here include projected sustainable transport in categories of rail, bus and active transport (i.e. cycleways and pedestrian paths and share ways).

Table 3.5 Value Capture Planning (VCP) Table: Public and Open Spaces – Waterloo Housing Estate Renewal Precinct

Step	1	2	3	4	5
Checklist	Public and open space types (Govt. land is 13 ha and private is 27 ha)	Meeting criteria for public and open spaces provision (Examples)	Projected new public and open spaces	Address design guidelines for public and open spaces (Examples)	Cost of public and open spaces provision by type
√	1. Natural areas	Natural area regeneration	ha	Amount of regeneration	$
√	2. Open spaces	Active and passive spaces	ha	Connecting public and open spaces	$
√	3..Public spaces	Need for public spaces	ha	Amount of spaces required	$
Totals			ha		X $

Steps as per columns	
1	Tick the checklist for public and open spaces types that apply
2	Adopt criteria for public and open spaces provision
3	Projected new public and open spaces
4	Refer to design guidelines for public and open spaces
5	Cost of public and open spaces provision by type (total cost shown as X$)

3.6.4 Developer Provisions and/or Levy Alternative

Having applied the value capture model to housing, open spaces and transport, the calculation of a value capture developer provision and/or levy alternative is addressed in Addendum 3.4 (Table 3.7).

3.7 Extending the VCP Model Beyond the Study Precinct

The application of the VCP model beyond the Waterloo Housing Estate Precinct can also be considered. This would enable the planning authorities (e.g. State and local councils) to consider applications to: the Central to Eveleigh Renewal Area (CERA); the City of Sydney overall; and, other inner city areas of Greater Sydney.

3.8 Results and Conclusions on Applying the Model

In applying the value capture planning (VCP) model (via the four tables) to the Waterloo Housing Estate Renewal Precinct there were a number of results and conclusions. Examining the first planning principle (*Equitable Housing*) (Table 3.4) it

Table 3.6 Value Capture Planning (VCP) Table: Sustainable Transport – Waterloo Housing Estate Renewal Precinct

Step	1	2	3	4	5
Checklist	Sustainable transport provisions	Meeting criteria for sustainable transport provisions (Examples)	Address design guidelines for sustainable transport (Examples)	Projected new transport provisions (i.e. infrastructure)	Cost of new transport provisions by type
✓	1. Rail (heavy and light)	Demand and access	New station, transport or interchange	State/Developer	$
✓	2. Bus	Demand and access	Bus transit way	State/Developer	$
✓	3. Active transport	Cycleways, Pedestrian paths, and share ways	Cycling and walking links	Council/ Developer	$
Total					Y $

Steps as per columns	
1	Tick the checklist for sustainable transport types that apply
2	Adopt criteria for sustainable transport provisions
3	Refer to design guidelines for sustainable transport provisions
4	Projected new transport provisions
5	Cost of new transport provisions by type (total cost shown as Y$)

Table 3.7 Value Capture Planning (VCP) Table: Developer Provision and/or Levy Alternative for Public and Open Spaces, and Transport Infrastructure – Waterloo Housing Estate Renewal Precinct

Step	1	2	
		Developer provision or levy alternative per new housing unit	
Checklist	Projected units of new housing by type and % (see Table 3.4)	Public and open spaces X $ (see Table 3.5)	Transport infrastructure Y $ (total) (see Table 3.6)
✓	All new housing (example) 7200 (100%)	**X $ + Y $** ÷ 7200 units = Z $ per unit	
✓	1. Affordable housing 1440 (20%)	Z $ per unit for 1440 units	
✓	2. Social housing 3600 (50%)	Z $ per unit for 3600 units	
✓	3. Market housing 2160 (30%)	Z $ per unit for 2160 units	

Steps as per columns	
1	Tick the checklist for housing types that apply and projected new housing by type and %
2	Value capture levy per new housing unit for public and open spaces, and sustainable transport infrastructure (Z $ per unit)

was shown that value capture can ensure public lands (such as this renewal area) can meet affordable, social and market housing needs. At the same time (it was indicated) the public (via the State) can benefit from land value increases in the rezoning and land use upgrading process of a renewal site. In addition housing is provided to balance the overall population mix of the upgraded estate (adding further land value to the site).

Moving to the second planning principle (*Public and Open Spaces*) within the model a checklist of steps to apply the model simplified the process (Table 3.5). Given the renewal site will cater for a very high density population, open spaces had been designed and proposed to the State. It was shown how the Sydney City Council significantly expanded these space (reference Sect. 3.5.4 above). As a redevelopment site the project could expand opportunities for connecting public and open spaces (including replicating natural area features lost over years).

Moving to the third planning principle (*Sustainable Transport*) within the model a further checklist was provided (Table 3.6). The development of the new Waterloo Metro rail station and associated residential towers and commercial service businesses would be major contributors to land value increases. These increases could flow partly back to the pubic under value capture (via the State ownership of most of the renewal site). Significant increases in State asset value and potential land leases and/or sales should create a significant revenue stream. Part of this stream could be applied to undertaking other transport initiatives (i.e. bus services and active transport) within the study area as noted in Table 3.6.

The forth part of the model to be applied was the developer provision (and/or levy alternative) (Table 3.7). The total cost of public and open spaces was added to the total cost of sustainable transport infrastructure to achieve an overall costing associated with the Waterloo Housing Estate Precinct development. To achieve a spread of the costings per housing unit this total cost was then divided by the number of development units by type (e.g. affordable, social and market). It was then shown that these costings could be a combination of developer provision and/or levy alternative per new housing unit.

In conclusion, in applying the model, there is an overall end result (in costing and meeting needs stemming from proposed development). It is shown that the tables can be used for individual planning principles or any combination of planning principles. It was also suggested that the application of the VCP model to larger areas beyond the Waterloo Housing Estate Precinct could also be considered. A comparison of these results with the results of applying the model to the three other study sites (Chaps. 4, 5, and 6) will be addressed in Chap. 7.

3.9 Summary

The chapter addressed renewal of inner city areas and applied a value capture planning (VCP) model. Two subjects (from earlier chapters) of significance to inner city areas were firstly summarised: (1) the implications of the coronavirus pandemic

(2020) for future urban planning (Chap. 1); and, (2) VCP model (including components of housing, public and open spaces, and sustainable transport) (Chap. 2). The Central to Eveleigh Renewal Area (CERA) (under the State of New South Wales and within the City of Sydney) was then introduced as the study area. The VCP model was then applied to a trial precinct area of the Waterloo Housing Estate Renewal Precinct within the CERA. Results and conclusions were then drawn, including the potential application of the model to other precincts. The application of the model to larger areas beyond the Waterloo Housing Estate Precinct was also examined. This included application to: the CERA, the City of Sydney overall; and, other inner city area of Greater Sydney. Finally, results and conclusions on applying the model were summarised.

Addendum 3.1 Timeline of Development and Aboriginal History in Redfern and Waterloo 1790 – 2000s

Year	Events
1790s	Site of today's Belmore Park and Central Station is the location of Aboriginal performances, ceremonies and trials.
1950s–1970s	Waterloo and Redfern public housing estates are constructed (included needed Aboriginal Housing).
1960s	Eveleigh workshops become obsolete. Aboriginal population exceeds 12,000. Redfern often referred to as a 'slum' due to unemployment, crime and negative media coverage. Many who can afford to leave move, as increasing numbers of underprivileged people move into the area.
1968	NSW Department of Housing begins resettling Aboriginal people in areas like Campbelltown and Mount Druitt.
1970s	The 1967 referendum and better social policies represent a key turning point for the community. The Aboriginal population grows to over 35,000 as many people relocate from rural areas. A number of community controlled services, such as legal, medical and children's services, are developed. The Aboriginal Medical Service becomes the first Aboriginal community-run medical service in Australia.
1973	The Australian Government buys a number of houses around Louis Street after a formal submission for funding from the area's community. This is the first housing collective in the country and the first successful land rights claim by an Aboriginal community.
1984	The Eora Centre is established in Redfern (now located in Darlington), as a community focused education centre for Aboriginal and Torres Strait Islander people.
1988	The University of Technology Sydney (off Broadway) is founded. The university continued to expand into the 2010's.
1992	In the United Nations Year for the World's Aboriginal Peoples, Paul Keating delivers an iconic speech on dispossession in Redfern Park.
1997	The Aboriginal Housing Company approves the demolition of 'The Block' and initiates the Pemulwuy Project.

(continued)

Year	Events
2000-2020	The rapid gentrification of the area brings about a significant shift in the demographic, placing renewed pressure on the Aboriginal and working class families that remain. An influx of students and young professionals sees the development of many cafés and bars, and the conversion of industrial spaces into studios and apartments. High rise development in Botany Rd and at Redfern Station continued to expand. The Aboriginal Housing Company developed final plans for The Block, as approved by the State following a planning panel review in March 2019 (under construction in 2020).

Source: Sydney City Council Archives 2020

Addendum 3.2 Central to Eveleigh Renewal Area (CERA) Key Planning Guidelines

1. Deliver Sustainable Transport

    1.1 Reduce car use by over 90% (when compared to the Sydney Metropolitan average of 20 km per person per day) through walking/cycling connections, better access to public transport, reduced parking rates and increased car-share options.
    1.2 Plan for the majority of residents to live within an 8 min walk to: heavy rail, a supermarket, community health, school/education facilities.
    1.3 All residents to live within 5 min or 400 m of public open space.

2. Increase energy and water efficiency and reduce waste in precinct planning

    2.1 Reduce total greenhouse gas emissions and resident water consumption
    2.2 Set BASIX (Building and Sustainability Index) and Green Star Ratings as targets for residential and commercial buildings
    2.3 Explore ways to maximise renewable energy opportunities
    2.4 Explore ways to support local food production
    2.5 Investigate how new buildings can have rainwater storage, access to low carbon energy sources
    2.6 Install live energy and water consumption monitors
    2.7 Use non-potable water supply to provide primary irrigation for all new public open spaces
    2.8 Not less than 50% of public open space to receive 4 h of direct sunlight in winter

3. Deliver biodiversity and environmental systems

    3.1 Increase the green canopy
    3.2 Ensure new plantings are endemic to Sydney and create diversity within plant families, genera and species

4. Adapt to climate change

   4.1 Mitigate heat island effects by reducing dark surfaces/hard surfaces in the built form and in the public domain, and by increasing canopy cover
   4.2 Through sustainable design, reduce household costs associated with housing, transport and utilities (compared to the Sydney metropolitan average) to help respond to changes in economic systems and energy prices

Addendum 3.3 State Engagement of the Community and Other Parties in Waterloo Renewal Planning

The State of NSW outlined (2017) its approach to the engagement of the community and other parties in the Waterloo renewal planning as follows.

Meeting Local Community Needs

> Our role on the project review panel will ensure we can make a significant contribution to the strategic planning process. We will aim to ensure urban renewal of Waterloo precinct meets the needs of the local community and is driven by community engagement and good design outcomes. The project can deliver a mix of public, affordable and private housing across the area to promote equity, social cohesion and inclusion. Essential support services need to be provided locally. The historical and cultural significance of Waterloo for our Aboriginal and Torres Strait Islander communities must be reflected in the precinct design and in the services and employment opportunities the redevelopment will provide. High quality open spaces and public areas of sufficient size must be incorporated into the precinct design to meet the needs of future residents and workers.

Social Housing

> The state government has committed to maintain the number of social housing dwellings in Waterloo precinct. We will continue to work collaboratively with the Department of Family and Community Services to: 1. support social housing tenants during the urban renewal process; 2. improve local amenity and safety; 3. enhance community wellbeing; and, 4. manage waste and cleansing. Our dedicated social housing project manager meets regularly with social housing tenants in the area and advocates on their behalf to address any issues of concern. The Lord Mayor hosts 2 open forums in Waterloo each year, bringing together senior staff from key agencies such as Family and Community Services, NSW Police and the local member of NSW Parliament. The City held a public meeting in June 2016 about the state government's plans for Waterloo. A follow-up meeting was held in August 2016. https://www.cityofsydney.nsw.gov.au/council/publications-updates/featured-articles/central-to-eveleigh-strategy

Advice for Tenants

> We have provided funding to Redfern Legal Centre to support social housing tenants affected by the redevelopment of the Waterloo estate. The centre's Waterloo Tenancy Advice Outreach Service provides free, independent advice and information about relocations and other tenancy-related issues. The service operates from the Factory Community Centre at 67 Raglan Street, Waterloo. You can also contact the Inner Sydney Tenants' Advice and Advocacy Service https://www.tenants.org.au/taas/istaas on 02 9696 5975. The City takes part in the Waterloo Redevelopment Group and other forums. The City funds the Waterloo Public Housing Action Group to run tenant-led consultation activities.

Addendum 3.4 Waterloo Housing Estate Renewal Precinct and Value Capture Planning (VCP) Tables

The Addendum applies (in Tables) three planning principles and components (in brackets) of those principles as follows (Chap. 2): (1) *Equitable Housing* (affordable, social and market). (2) *Public and Open Spaces* (natural areas, open spaces, and public spaces); and, (3) *Sustainable Transport* (rail, bus and active transport). In addition, the second part of the model incorporating *Developer Provisions and/or Levy Alternative* is applied within a table. These tables will be applied to the Waterloo Housing Estate Renewal Precinct. The tables enable an authority (i.e. State and municipal) to calculate needs stemming from new housing developments. The tables (in above order) allow calculation of: (1) projected units of new housing; (2) projected public and open spaces and costing; (3) projected sustainable transport provisions and costing; and, (4) developer provisions and/or levy alternative. The planning principle Equitable Housing is illustrated first.

Equitable Housing

The first planning principle (equitable housing) in Table 3.4 incorporates four steps for the planning body to follow in application to Waterloo Housing Estate Renewal Precinct.

A checklist is provided to identify housing types (Step 1). Step 2 is meeting criteria of need for housing types. Step 3 is addressing design guidelines for housing types. The projected units of new housing by types (Step 4) can be calculated by number and %. The planning body can then use these numbers within developer agreements (Table 3.7).

Public and Open Spaces

The second planning principle (public and open spaces) (Table 3.5) incorporates five steps for the planning body to follow in application to Waterloo Housing Estate Renewal Precinct.

A checklist is provided to identify public and open spaces types (Step 1). Step 2 is meeting criteria of need for public and open spaces provision. Step 3 is projecting new public and open spaces by type. Step 4 is addressing design guidelines for public and open spaces. Step 5 calculates the cost of the new public and open spaces. The planning body can they use these numbers within developer agreements or levy alternatives (Table 3.7).

Sustainable Transport

The third planning principle (sustainable transport) (Table 3.6) incorporates five steps for the planning body to follow in application to Waterloo Housing Estate Renewal Precinct.

A checklist is provided to identify sustainable transport provisions (Step 1). Step 2 is meeting criteria for sustainable transport provisions. Step 3 is addressing design guidelines for sustainable transport. Step 4 is projecting sustainable transport provisions for new public by type. Step 5 calculates the cost of new transport provisions. The planning body can they use these numbers within developer agreements or levy alternatives (Table 3.7).

Developer Provision and/or Levy Alternative

The forth planning principle (developer provision and/or levy alternative for public and open spaces) (Table 3.7) incorporates two steps for the planning body to follow in application to Waterloo Housing Estate Renewal Precinct.

Application

A checklist is provided to address the projected new housing (Step 1) by type and % from Table 3.4. Step 2 identifies the developer provision or levy alternative per new housing unit for public and open spaces (Table 3.5) and sustainable transport provisions (Table 3.6). The user of the table can then use these numbers in formulating development agreements and/or levy alternatives among stakeholders.

References

NSW Government (2017) Central to Eveleigh urban transformation strategy. NSW Government, Sydney

NSW Government (2018) Greater Sydney region plan: a metropolis of three cities. NSW Government, Sydney

Part III
Middle Ring City Renewal and Value Capture Planning

Part III addresses middle ring city renewal and value capture planning (VCP) (outlined in Chap. 1). The VCP model (developed in Chap. 2) is applied to Greater Sydney's Sydenham to Bankstown Corridor (SBC). The Canterbury Renewal Precinct (within the City of Canterbury Bankstown) is taken as a case study within the Corridor.

Chapter 4
Middle Ring City Renewal and Value Capture Planning – Greater Sydney's Sydenham to Bankstown Urban Renewal Corridor (SBURC)

Abstract This chapter addresses renewal of middle ring city areas and applies a value capture planning (VCP) model. Two subjects (from earlier chapters) of significance to middle ring city areas are firstly summarised: (1) the implications of the coronavirus pandemic (2020) for future urban planning (Chap. 1); and, (2) VCP model (including planning principles of equitable housing, public and open spaces, and sustainable transport) (Chap. 2). The Sydenham to Bankstown Urban Renewal Area (SBURA) (under the State of New South Wales and mostly within the City of Canterbury Bankstown) is then introduced as the study area. This is the second of four selected study areas. The VCP model is then applied to the case study Canterbury Renewal Precinct, within the SBURA. Results and conclusions are then drawn, including the potential application of the model to other precincts within the study area. The application of the model to larger areas beyond the Canterbury Renewal Precinct is also addressed, including application to: the City of Canterbury Bankstown; the SBURA; and, other Greater Sydney middle ring areas. At chapter's end it's noted that Chap. 7 will address: (1) overall results of applying the model to each of the four case study areas; and, (2) the application of the model to any urban or regional centre elsewhere in Australia or in the world.

Dedication There are many people in the history of cities and renewal who have left major legacies. This chapter is dedicated to two of them (Plate 4.1), Sir Joseph Banks (naturalist and botanist) and Kevin Moss (Parliamentarian NSW)

Sir Joseph Banks, 1st Baronet (1743–1820) was an English naturalist, botanist, and patron of the natural sciences. He took part in Captain James Cook's first great voyage (1768–1771). He advised King George III on the Royal Botanic Gardens, Kew, and by sending botanists around the world to collect plants, he made Kew the world's leading botanical gardens. Banks advocated British settlement in New South Wales and colonisation of Australia. He is credited with introducing the eucalyptus, acacia, and the genus named after him, *Banksia*, to the Western world. Around 80 species of plants bear his name. (Source: Wikipedia 2020)

Kevin Joseph Moss (1946-) is an Australian politician. He was a Labor Party member of the New South Wales Legislative Assembly from 1986 to 2003, representing the electorate of Canterbury. Elected as Canterbury Council's Mayor in 1980. He was re-elected four times as the member for Canterbury, and served as a parliamentary secretary in the first two terms of the Carr government. (Source: Wikipedia 2020)

R. Rauscher, *Renewing Cities with Value Capture Planning*,
https://doi.org/10.1007/978-3-030-62958-8_4

Plate 4.1 Dedications: Sir Joseph Banks and Kevin Moss

4.1 Middle Ring City Renewal and Value Capture Planning (VCP)

The Greater Sydney's City of Canterbury Bankstown is a middle ring city and selected in Chap. 2 as a case study to apply a value capture planning (VCP) model to. As noted in that chapter the implications of the coronavirus pandemic (2020) (Chap. 1) on the City's planning is important to examine first. One urban planning implication from the coronavirus impact is the likely overall review of Canterbury Bankstown planning. This would include: examining residential densities and settlement patterns; transport; unban infrastructure provisions; and, public and open space needs. VCP could thus become an essential and pivotal planning tool within urban development policy making in a post coronavirus era. Greater Sydney and the City of Canterbury Bankstown are introduced next.

4.2 Greater Sydney and City of Canterbury Bankstown City

The City of Canterbury Bankstown (Plate 4.2) is in Greater Sydney's south west (Canterbury being 12.5 km and Bankstown 29.3 km respectively from the Sydney GPO). The City population of 388,000 (2020) is part of Greater Sydney's 5 m residents (2020). The district level planning affecting the City of Canterbury Bankstown (as with other districts of Greater Sydney) comes under the New South Wales (NSW) Dept. of Planning, Industry and Environment (DPIE) (as outlined in Chaps. 1 and 2). The City development is examined next.

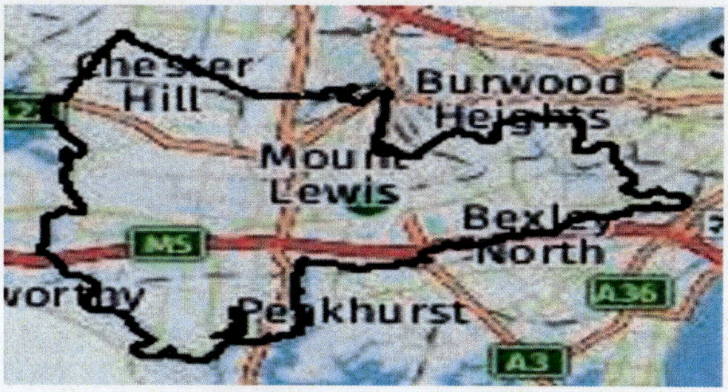

Plate 4.2 Canterbury Bankstown City boundaries. (Source: Gregory's Maps 2020)

Plate 4.3 Development concepts at Bankstown. *Left to right:* model of Bankstown development; a high rise residential project; and, an illustration of proposed upgraded Bankstown rail station. (Source: Sydney Morning Herald 2014)

4.3 Canterbury Bankstown City Development

To gain a perspective on the New South Wales (NSW) plans (Chap. 2) plans for the City of Canterbury Bankstown, it's useful to look at recent developments in the main centres of the city. There has been considerable medium and high density residential building in the two districts of the city, Bankstown and Canterbury, over the last couple of decades (and continues in 2020). Looking at Bankstown centre first, development concepts are illustrated in Plate 4.3. This includes (left to right): model of Bankstown centre development; example of residential high rise development; and, artist's sketch of entry to proposed upgraded Bankstown station (under the Sydney Metro South West, addressed later in Chapter).

From development concepts to practice, medium rise (built or proposed) development at Bankstown is illustrated in Plate 4.4. This reflects several thousand units built over the last few decades (1980's to 2020) at Bankstown centre. The attraction (as advertised by developers) to these centres, is: (1) locational advantage to shops, eateries, services rail/bus (just over ½ h to Sydney CBD); and, (2) road access to a major road (M5 motorway), schools and cultural facilities.

Plate 4.4 Medium-rise residential built or proposed at Bankstown. (Source: Build Sydney July 2020)

From medium-rise residential to high-rise, Plate 4.5 illustrates examples of high-rise development (built or proposed) at Bankstown Centre. These buildings represent an overall trend to high-rise renewal and apartment living in the centre (thus small family sizes).

Moving from Bankstown to Canterbury centre, examples of medium density and high-rise development (built or proposed) are illustrated in Plate 4.6.

Looking now at future streetscapes in Canterbury Bankstown City centres, an artist impression of medium and high rise development in the Canterbury district is noted in Plate 4.7.

The Sydenham to Bankstown Corridor (https://www.cbcity.nsw.gov.au/council/planning-for-the-city/sydenham-to-bankstown-corridor) has been the focus of significant NSW Government-led renewal planning. Within the draft Sydenham to Bankstown Urban Renewal the State plans for residences to be within walking distance of the proposed metro stations between Sydenham and Bankstown (Plate 4.8). The Sydenham to Bankstown urban renewal corridor is examined next.

Plate 4.5 High-rise residential built or proposed at Bankstown Centre. (Source: Build Sydney 26 Oct 2017)

4.4 Sydenham to Bankstown Urban Renewal Corridor (SBURC)

The State adopted the Sydenham to Bankstown Urban Renewal Corridor (SBURC) to guide development within the corridor (as noted above). Land uses proposed in rezoning hubs along the Corridor are shown in Plate 4.9. Canterbury is located in the centre, with Hurlstone Park, Dulwich Hill, and Marrickville to the right and Campsie and Belmore to the left. These centres are all urban renewal precincts under the Corridor plans.

Given the above projected land uses and housing (incorporated within the State Government's renewal planning), key planning factors to be examined next include: (1) Canterbury and Campsie Locality; (2) Sydney Metro Southwest; (3) Planning Proposals and Issues; (4) State, Council and Community Engagement; (5) Priority Precincts and State Infrastructure; and, (6) Housing Urban Design in the Corridor.

4.4.1 Canterbury and Campsie Locality

Canterbury and Campsie locality contains two major hubs, Canterbury and Campsie centres, examined now in more detail. The renewal plans for Canterbury and Campsie locality come under the Sydenham to Bankstown Renewal Corridor

Plate 4.6 High-rise residential built and proposed at Canterbury. (Source: Build Sydney July 2020)

Plate 4.7 Canterbury district illustrations of centres' upgrades. *Left to right:* Campsie's main street; Campsie square; Belmore priority precinct; and park, upgrading at Campsie. (Source: Sydney Morning Herald 2013)

(SBURC) (Table 4.1 below). The Dept. of Planning, Industry and Environment (DPIE) notes that its planning includes community's input and Councils' vision to underpin planning of the area. The planning is directly affected (being central in the Corridor) by the Sydney Metro Southwest (referenced in table below).

Plate 4.8 Sydenham to Bankstown proposed open spaces at proposed Metro Station. (Source: NSW Government)

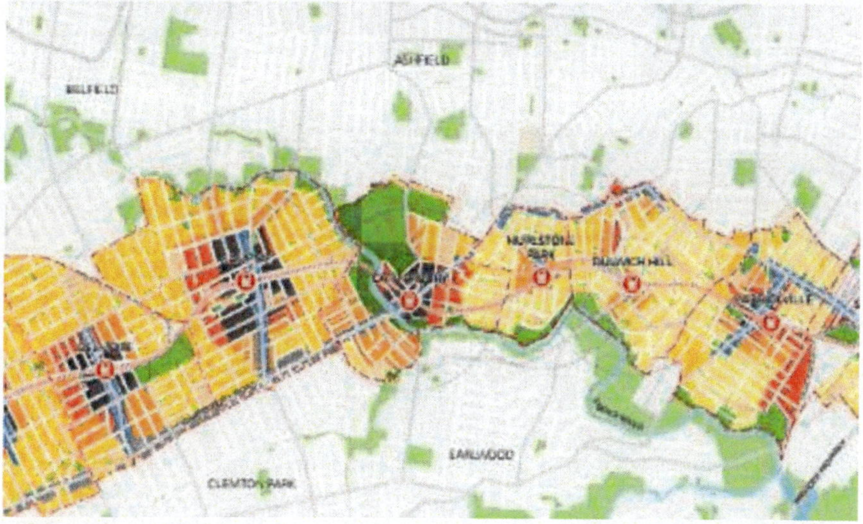

Plate 4.9 Land uses proposed in rezoning hubs in Sydenham to Bankstown Corridor. (Source: Sydney Morning Herald 2014)

Table 4.1 Canterbury and Campsie Locality and Sydenham to Bankstown Urban Renewal Corridor (SBURC)

Renewal plan	Aspects of plan	Web access
Canterbury and Campsie Locality	Precincts Renewal	https://www.planning.nsw.gov.au/Plans-for-your-area/Priority-Growth-Areas-and-Precincts/Sydenham-to-Bankstown-Urban-Renewal-Corridor
Metro Sydney South West	Upgrade of rail to new Metro system	https://www.sydneymetro.info/citysouthwest/environment-planning

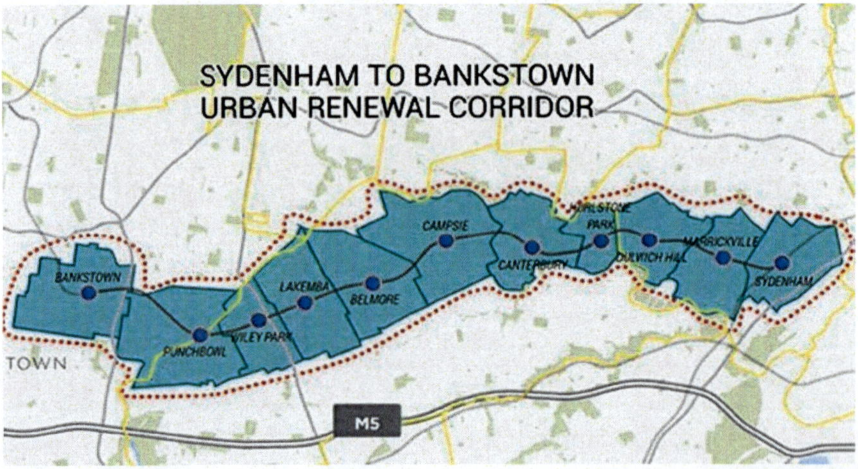

Plate 4.10 Sydenham to Bankstown Urban Renewal Corridor. (Source: NSW Government 2020)

4.4.2 *Sydney Metro Southwest*

Canterbury and Campsie are both affected by the Sydenham to Bankstown (Sydney Metro Southwest) (SMS) (stretching 24 km Sydney CBD to Bankstown). The SMS will see 11 stations noted upgraded (construction is expected to be completed in 2024). The rail project, which covers 13 km of the existing Bankstown rail line (running from Bankstown in the south west to Sydenham in north east direction), is shown within the corridor (Plate 4.10).

Railway stations within and outside the corridor are the urban hubs as shown in Plate 4.11.

In addition, Plate 4.12 provides a closer look at the Sydenham to Bankstown Urban Renewal Corridor renewal area and individual precincts along a section of the Metro line (between Sydenham and Dulwich Hill).

Upgrading of public areas and introduction for new trains (driverless) on the Sydney Metro Southwest (Plate 4.13) will create a new express service run. The plate shows the Sydenham Station's proposed new public area and a design of the new metro train (right).

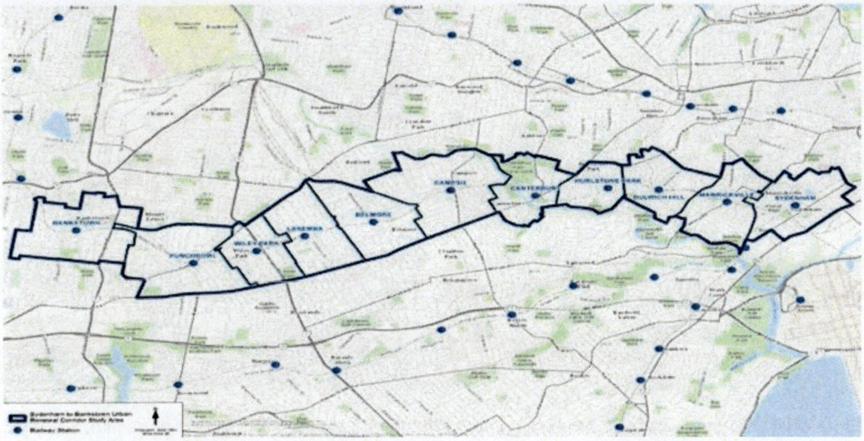

Plate 4.11 Sydenham to Bankstown Urban Renewal Corridor rail stations in corridor and outside. (Source: NSW DPIE 2020)

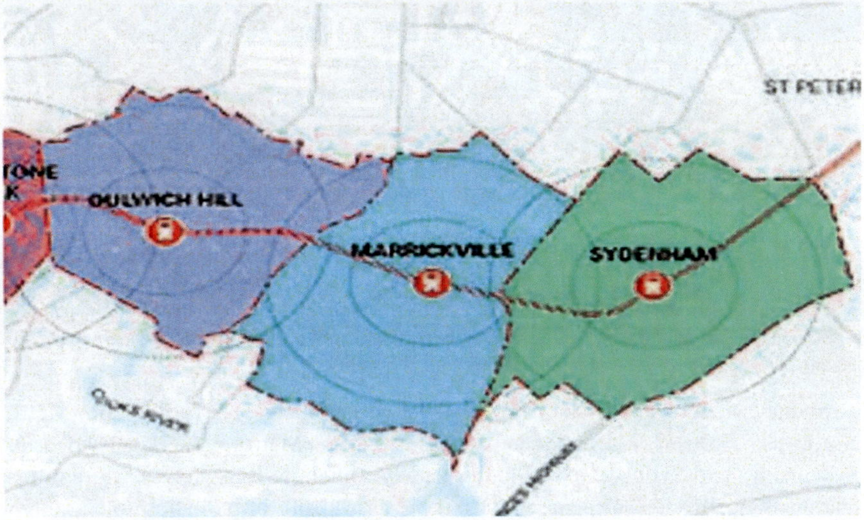

Plate 4.12 Sydenham to Dulwich Hill section of Bankstown Urban Renewal Corridor. (Source: NSW Government 2020)

Further rail station upgrades as proposed in the SMS are illustrated in Plate 4.14. A platform area is shown on the left. A concourse is shown in the centre frame. In addition, a new train for the Metro line is shown on the right.

In outlining the SMS line the NSW Planning Minister (Rob Stokes) (2017) noted:

> The eleven stations will be converted to a single deck Metro line and connected to Chatswood via a new Harbor crossing. The Metro project is expected to be completed by 2024 (Rouse Hill to Bankstown). The Sydney Metro will provide up to 15 trains per hour

Plate 4.13 Sydney Metro Links. *Left to right* Sydenham proposed new public area; and, a design of the new metro train. (Infrastructure NSW 2020)

Plate 4.14 Sydney Metro Southwest Project. (Source: NSW State)

between Bankstown and Sydenham and reduced travel times of up to 10 minutes. The plan also says there is potential for light rail between Bankstown and Parramatta via Granville and Chester Hill. (Source: Planning Minister's press release 2017)

Sydney Metro is a new, standalone rail network, which consists of Sydney Metro Northwest (Rouse Hill and Chatswood), Sydney Metro City (Chatswood to Sydenham), and Sydney Metro Southwest (Sydenham to Bankstown). More information is on the Department's Major Projects website http://majorprojects.planning.nsw.gov.au/index.pl?action=view_job&job_id=7400

The Sydenham to Bankstown upgrade is a program of works to upgrade 13 km of the Sydney Trains T3 Bankstown Line and convert 10 stations between Sydenham and Bankstown to metro standards. Sydney Metro's Preferred Infrastructure Report (SPIR) (2017) provides a description of changes made to the proposal exhibited in the Environmental Impact Statement (EIS) (Sydenham to Bankstown Upgrade). Sydney Metro Southwest is addressed in more detail next.

4.4.3 Planning Proposals and Issues

There are many planning issues raised by communities in respect to development proposals under the aegis of the State or local authority. The NSW Dept. of Planning, Industry and Environment (DPIE) noted key issues raised during the workshops included: (1) the need to work within the local character of each area; (2) the need for good design, including sensitive building height transition controls; (3) the need for infrastructure and open space to support current and future communities; (4) the need to address future traffic increases and congestion; and, (5) a desire for improved connections within and across precincts.

The State noted that the draft strategy was being developed to harness the benefits of the Sydney Metro City and SMS projects. This includes upgrading the stations along the Sydenham to Bankstown corridor. A major community workshop report on the subject can be viewed at https://www.planning.nsw.gov.au/Plans-for-your-area/Priority-Growth-Areas-and-Precincts/Sydenham-to-Bankstown-Urban-Renewal-Corridor/Documents

The Canterbury Bankstown Council (representing its constituents) also commented on the Metro Upgrade project along the Sydenham to Bankstown line. Council comments (2018) are found at https://www.cbcity.nsw.gov.au/council/planning-for-the-city/sydenham-to-bankstown-corridor/sydenham-to-bankstown-metro-upgrade-project

Subsequently in late 2018, Council made a **submission** to DPIE in response to the Preferred Infrastructure Report for the project. The submissions set out why Council could not support the Metro Southwest concept in its proposed form. Council's submission stated that the community deserves transport infrastructure that enhances liveability, amenity and economic prosperity. The report contact details for related matters can be found at the following links:

Metro SW and Keeping in Touch
https://www.sydneymetro.info/get-touch

Metro SW and Overview
https://www.sydneymetro.info/citysouthwest/project-overvie

Metro SW and News
https://www.sydneymetro.info/news-article-category/city-southwest

Metro SW and Planning
https://www.sydneymetro.info/citysouthwest/environment-planning

Metro SW and Other NSW Infrastructure Projects
http://majorprojects.planning.nsw.gov.au/index.pl?action=view_job&job_id=8256

Given the above planning proposals and implications for value capture planning (VCP), the State, council and community engagement process is examined next.

4.4.4 State, Council and Community Engagement

The Canterbury Bankstown Council and many community groups within the proposed renewal area have questioned the State Government's planning of this Sydenham to Bankstown urban renewal corridor. The revised draft Sydenham to Bankstown Urban Renewal Corridor Strategy (2017) was exhibited from June to September 2017. More than 2800 submissions were received raising a range of issues, including infrastructure, open space, employment lands, affordable housing, transport and traffic, density, urban design, heritage and local character. The Department indicated that people who shared their views, included: 600 community members who attended six community consultation events; 900 survey respondents; and, more than 2800 submitters. Sydney Metro City and SMS are examined next as it will affect all precincts in the SBURC.

Canterbury Bankstown Council commented in 2019 that it could not support the Sydenham to Bankstown Corridor, submission within web site as follows: https://www.cbcity.nsw.gov.au/council/planning-for-the-city/sydenham-to-bankstown-corridor/sydenham-to-bankstown-urban-renewal-corridor-strategy

> Council's overall position is the Draft Strategy and Metro Environmental Impact Statement (Metro EIS) cannot be supported at this time. The Draft Sydenham to Bankstown Urban Renewal Corridor Strategy is narrowly focussed on delivering housing, without considering the capacity of the existing and proposed infrastructure required to support it, such as new hospitals and schools. Critically, the Draft Strategy is inconsistent with the Greater Sydney Commission's Draft Greater Sydney Region Plan and Draft South District Plan. The Metro proposal, as it currently stands, lacks the vision and commitment needed to provide economic, social and infrastructure opportunities for the people of Canterbury-Bankstown. Whilst the provision of much needed infrastructure to service the growing needs of our community is welcomed, Council has significant concerns the project, as planned, will have detrimental outcomes for our community and key opportunities are not being realised.

Council's position was that the Draft Urban Renewal Corridor Strategy contained a significant number of unresolved issues, particularly concerning infrastructure, open space, urban design and funding. In addition, there is a summary of main points of the Council's submission to the Draft Sydenham to Bankstown Urban Renewal Corridor Strategy at Submission to the Draft Sydenham to Bankstown Urban Renewal Corridor Strategy

Council's issues with the proposed infrastructure projects are contained in Attachment A (follows) and community issues as contained in Attachment B (also follows).

Attachment A – Council's issues with the proposed infrastructure projects
Attachment B – Community Issues Report

The State also noted that the Council's and community's comments about insufficient open space in parts of the Sydenham to Bankstown corridor. Some of the open space plans for the corridor included: (1) A potential 13 km expansion of the Greenway to Bankstown should surplus rail land become available; (2) Providing public access to school playing fields after school hours; (3) Improving existing open space in the area; (4) The Special Infrastructure Contribution (SIC) Plan would include an allocation to improve district level open space in the corridor such as Tasker Park and Terry Lamb Reserve; (5) Providing open space in the redevelopment of large sites such as the Carrington Road Precinct; and (6) Identifying Priority Precincts at Campsie, Canterbury, Belmore and Lakemba to ensure the delivery of new open space. The State government adopted a new 'priority precincts' program, which would affect the Sydenham-Bankstown corridor (examined next).

4.4.5 Priority Precincts and State Infrastructure

The State announced that Canterbury, Campsie and Belmore were to be nominated as priority precincts for State Government. The government's Priority Precinct Program was to be included in its 20-year strategy for the Sydney Metropolitan Area, *A Plan for Growing Sydney* (2014). The program's aim was to boost housing supply in designated infill areas by revitalising local centres and services. Each nominated precinct was to receive $5 million for community infrastructure, on top of developer levies from Section 94 plans. Canterbury, Campsie and Belmore are expected to develop first among the seven suburbs along the rail corridor. This would involve more high density around the suburbs transport nodes. This would mainly be around the State's future metro stations. It would be part of the Sydney Metro Stage 2 project which would see a conversion of the Bankstown line from double decker trains to metro trains. There will be trains running every 4 min in the peak which will be able to serve 1000s of extra passengers per hour. This would ensure faster journey times into the city thus making it a great location to increase density. How will infrastructure be planned and funded?

With increased population and more housing, infrastructure and open space would be needed. The Planning Minister indicated that when more significant State infrastructure, such as emergency services, schools, state and district roads, district open space and transport facilities, were needed. He noted the State would consider declaring the precinct a Special Contribution Area. This means infrastructure will be funded by developers through a Special Infrastructure Contribution (SIC). The benefits to the community (he indicted) included: living and working near public transport, shops and services. Finally, he indicated that having a range of housing also helps people live close to family and friends, no matter what their life stage. Moving to planning on the ground, housing and design factors within corridor planning are examined next.

4.4.6 *Housing and Urban Design in the Corridor*

Moving from the Metro project to housing and design a background on the issue is important for value capture planning (VCP). This is the first model principle (*Equitable Housing*) within the Sydenham to Bankstown Urban Renewal Corridor (SBURC). The State notes that the plans call for new homes to be within a 5–10-min walk of the new Sydney Metro (https://www.sydneymetro.info/citysouthwest/project-overview). The government also notes that the homes will be spread along a 13.5 km, 11 metro station corridor. The government acknowledges that the next phase of the planning process will shape what the precincts will look like and what services will be are required. This includes cafes, shops, open space, transport and education.

The State estimates that over the next two decades, Greater Sydney will need a minimum 725,000 new homes (that's over 36,000 each year) as noted earlier. This would allow for a Greater Sydney expanding population predicted to grow by an extra 2.1 million people to 6.4 million by 2036. The Department states that it plans to do this without compromising the valuable local character and heritage of the precincts, thus stating the need for good design.

On that design side the Government stated that it used the expertise of NSW Government Architect. The State refers to its design guidelines under Better Placed (Government Architect 2020) https://www.governmentarchitect.nsw.gov.au/policies/better-placed. These design guidelines include sensitive building height controls, the need for essential infrastructure and adequate open space.

This plan is a guide to ensure good design outcomes in developing other plans. This means (the Government notes) these precincts will need to be well designed, sustainable, adaptable, connected, safe and comfortable for residents to examine. Heritage items, which contribute so much to the character of an area (State notes) will continue to be protected and considered in the precinct plans. Finally, the State included the Canterbury Racecourse (a major site) in the revised strategy. The government recommended the racecourse be investigated for new homes and open space. The Canterbury Renewal Precinct is examined next.

4.5 Canterbury Renewal Precinct Case Study

The Canterbury suburb is shown outlined in Plate 4.15.

The Canterbury Renewal Precinct case study will be examined now in more detail.

Zoning of the Canterbury Renewal Precinct (at the core centre of the suburb) is illustrated in Plate 4.16.

An aerial view shows the core of the Canterbury Renewal Precinct (study area) is shown here (Plate 4.17).

Plate 4.18 contains a sample of the author's survey showing views of the surrounding open spaces and waterway (Cooks River) of the Precinct. From left to

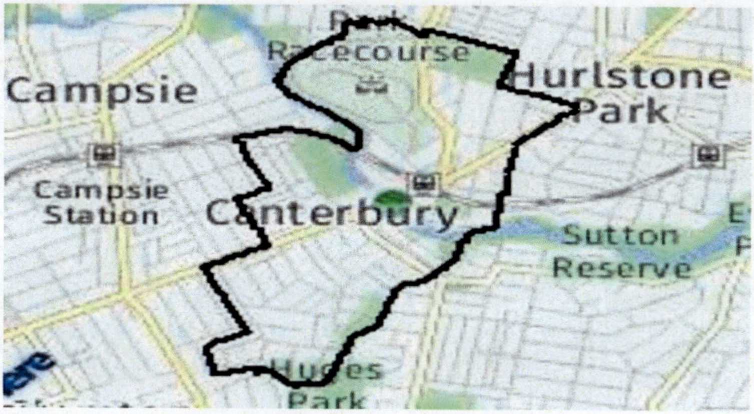

Plate 4.15 Canterbury suburb boundaries. (Source: State NSW)

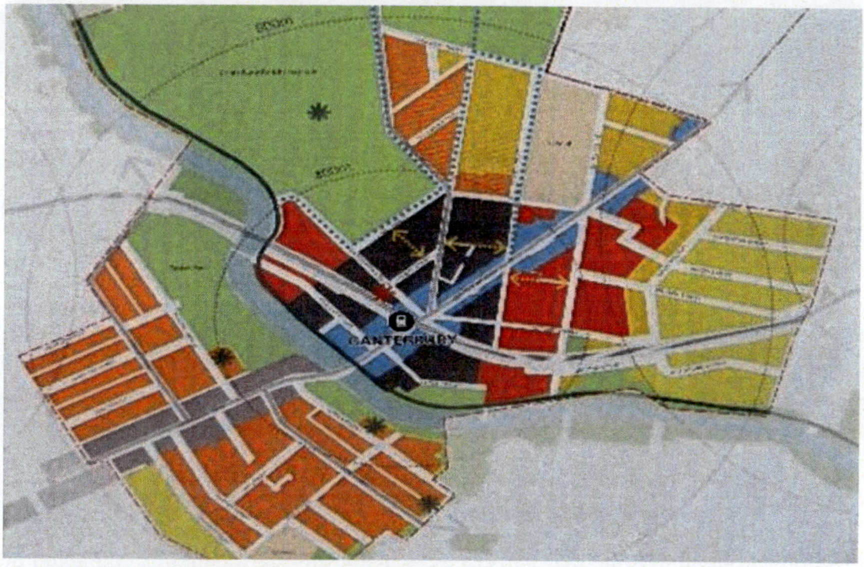

Plate 4.16 Zoning of Canterbury Precinct. (Source: NSW Government 2020)

right are: bridge over a section of the Cooks River; parklands with Precinct new high-rise in distance; and, pathway through restored foreshore of the Cooks River. The value capture planning (VCP) model is applied next to the Canterbury Precinct.

Plate 4.17 Canterbury from the air. (Source: State NSW 2020)

Plate 4.18 Canterbury Precinct Renewal Area open spaces. (Source: Ray Rauscher 2020)

4.6 Value Capture Planning (VCP) Model Application

The Value Capture Planning (VCP) model (Tables 2.1, 2.2, 2.3, and 2.4 in Chap. 2) is now examined in detail and applied to the Canterbury Renewal Precinct. Additional planning principles and components of those principles as contained in the VCP model can be added to or modified to reflect circumstances of any precinct. This is usually done as a collaborative exercise with all stakeholders (state planning authority, local council, development industry and community). The three VCP planning principles (applied within tables) as addressed in Chap. 2 are: *Equitable Housing*, *Public and Open Spaces*, and *Sustainable Transport*. In addition, the second part of the model incorporating *Developer Provisions and/or Levy Alternative* is applied within a table. Each of these can now be examined in detail.

4.6.1 Equitable Housing

The *Equitable Housing* principle (with components of affordable, social and market housing) of the value capture model for the Canterbury renewal area is illustrated Addendum 4.1 (Table 4.2). Key factors here include: projected units of new housing by type, number and % of all renewal housing: affordable at 30%; social at 20%; and, market at 50%.

4.6.2 Public Spaces and Open Spaces

The *Public and Open Spaces* principles (with components) of the value capture model for the Canterbury Renewal Precinct is illustrated Addendum 4.1 (Table 4.3). Key factors here include projected public and open spaces in categories of: natural areas; open spaces; and, public areas (quantities to be determined by final plans).

4.6.3 Sustainable Transport

The *Sustainable Transport* component (with elements of rail, bus and active transport) of the value capture model for the Canterbury renewal area is illustrated Addendum 4.1 (Table 4.4). Key factors here include projected sustainable transport

Table 4.2 Value Capture Planning (VCP) Table: Equitable Housing – Canterbury Renewal Precinct

Step	1	2	3	4
Checklist	Housing types	Meet criteria of need for housing types (Examples)	Addressing design guidelines for housing types (Examples)	Projected # and % of new housing units (Example)
	All new housing	Housing standards	Variety of designs	5000 (100%)
✓	1. Affordable housing	Special housing groups	Innovative architecture	1500 (30%)
✓	2. Social housing	Social support needs	Maintenance efficiencies	1000 (20%)
✓	3. Market housing	Housing as income compatible	Range of housing style choice	2500 (50%)

Steps as per columns	
1	Tick the checklist for housing types that apply
2	Adopt criteria for housing provision
3	Refer to design guidelines for housing
4	Projected units of new housing and % by type

Table 4.3 Value Capture Planning (VCP) Table: Public and Open Spaces – Canterbury Renewal Precinct

Step	1	2	3	4	5
Checklist	Public and open space types	Meeting criteria for public and open spaces provision (Examples)	Projected new public and open spaces	Address design guidelines for public and open spaces (Examples)	Cost of public and open spaces provision by type
✓	1. Natural areas	Natural area regeneration	ha	Amount of regeneration	$
✓	2. Open spaces	Active and passive spaces	ha	Connecting public and open spaces	$
✓	3. Public spaces	Need for public spaces	ha	Amount of spaces required	$
Totals			ha		X $

Steps as per columns	
1	Tick the checklist for public and open spaces types that apply
2	Adopt criteria for public and open spaces provision
3	Projected new public and open spaces
4	Refer to design guidelines for public and open spaces
5	Cost of public and open spaces provision by type (total cost shown as X$)

Table 4.4 Value Capture Planning (VCP) Table: Sustainable Transport – Canterbury Renewal Precinct

Step	1	2	3	4	5
Checklist	Sustainable transport provisions	Meeting criteria for sustainable transport provisions (Examples)	Address design guidelines for sustainable transport (Examples)	Projected new transport provisions (i.e. infrastructure)	Cost of new transport provisions by type
✓	1. Rail (heavy and light)	Demand and access	New station, transport or interchange	State/Developer	$
✓	2. Bus	Demand and access	Bus transit way	State/Developer	$
✓	3. Active transport	Cycleways, Pedestrian paths, and share ways	Cycling and walking links	Council/ Developer	$
Total					Y $

Steps as per columns	
1	Tick the checklist for sustainable transport types that apply
2	Adopt criteria for sustainable transport provisions
3	Refer to design guidelines for sustainable transport provisions
4	Projected new transport provisions
5	Cost of new transport provisions by type (total cost shown as Y$)

in categories of rail, bus and active transport (i.e. cycleways, pedestrian paths and share ways).

4.6.4 *Developer Provision and/or Levy Alternative*

Having applied the value capture planning (VCP) model to housing, public and open spaces and transport, consideration of developer provisions and/or levy alternative can be completed. Table 4.5 (Addendum 4.1) incorporates the steps for the State and/or local authority to apply under value capture planning (VCP) and for calculating that developer provision and/or levy alternative (per new housing unit).

4.7 Extending the VCP Model Beyond the Precinct

The application of the VCP model beyond the Canterbury Renewal Precinct can also be considered. This would enable the planning authorities (e.g. State and local councils) to consider applications to: the City of Canterbury Bankstown; Sydenham to Bankstown Urban Renewal Area (SBURA); and, other Greater Sydney middle ring areas.

Table 4.5 Value Capture Planning (VCP) Table: Developer Provision and/or Levy Alternative for Public and Open Spaces, and Transport Infrastructure – Canterbury Renewal Precinct

Step	1	2	
		Developer provision or levy alternative per new housing unit	
Checklist	Projected units of new housing by type and % (see Table 5.1)	Public and open spaces X$ (see Table 4.3)	Transport provision Y $ (total) (see Table 4.4)
✓	All new housing (example) 5000 (100%)	**X $ + Y $** ÷ 5000 units = Z $ per unit	
✓	1. Affordable housing 1500 (30%)	Z $ per unit for 1500 units	
✓	2. Social housing 1000 (20%)	Z $ per unit for 1000 units	
✓	3. Market housing 2500 (50%)	Z $ per unit for 2500 units	
Steps as per columns			
1	Tick the checklist for housing types that apply and projected new housing by type and %		
2	Value capture levy per new housing unit for public and open spaces, and transport infrastructure (Z $ per unit)		

4.8 Results and Conclusions on Applying the Model

In applying the value capture planning (VCP) model (via the four tables) to the Canterbury Renewal Precinct there were a number of results and conclusions. Examining the first planning principle within the model (*Equitable Housing*) it was shown that value capture could ensure renewal lands (e.g. within the Precinct) can meet affordable, social and market housing needs. At the same time it was shown that the public (via the State) could benefit from land value increases in the rezoning and land upgrading processes within a renewal area.

Moving to the second planning principle (*Public and Open Spaces*) it was shown how public and open space needs can to be designed to incorporate value capture planning. It was illustrated how renewal of the Canterbury Renewal Precinct lent itself to opportunities for new public and open spaces. This would include, for example, restoring natural areas along the Cooks River. These areas along the foreshores have been impacted over the years of development. Finally, it was noted that the land value increases from the precinct's development would enable public and open space provisions to be achieved.

Moving to the third planning principle (*Sustainable Transport*), the focus was on the upgrading of Canterbury rail station (under the Metro South West project). It was shown that this upgrading had a catalytic development effect on new residential and commercial businesses, including land value increases. These increases could be configured to flow partly back to the pubic (i.e. the State ownership of the station's site and curtilage). Significant increases in State asset value and potential land leases and/or sales could create a further revenue stream of public benefit. Part of this stream could be applied to undertaking further transport initiatives (i.e. bus service facilities, transit ways and active transport) as noted in Table 4.4.

Finally, moving to the developer it was shown how the developer provisions and/or levy alternative per new housing unit could be incorporated within the model. The total cost of public and open spaces was added to the total cost of sustainable transport infrastructure to achieve an overall costing. This total cost was then divided by the anticipated number of new units (e.g. affordable, social and market housing) to achieve a costing per housing unit.

In conclusion, there is a cumulative end result in applying the VCP model to the Canterbury Renewal Precinct (i.e. in costing and meeting public needs). It was shown that the model could be designed for one or more of the three planning principles (additional principles can be added dependent of circumstances of precinct renewal). It was also suggested that the VCP model application to areas beyond the Canterbury Renewal Precinct was feasible. A comparison of these results with the results of applying the model to the three other case study sites (Chaps. 3, 5, and 6) will be addressed in Chap. 7.

4.9 Summary

This chapter addressed renewal of middle ring city areas and applied a value capture planning (VCP) model. Two subjects (from earlier chapters) of significance to middle ring city areas were firstly summarised: (1) the implications of the coronavirus pandemic (2020) for future urban planning (Chap. 1); and, (2) VCP model (including planning principles of equitable housing, public and open spaces, and sustainable transport) (Chap. 2). The Sydenham to Bankstown Urban Renewal Area (SBURA) (under the State of New South Wales and mostly within the City of Canterbury Bankstown) was then introduced as the study area. The VCP model was then applied to the Canterbury Renewal Precinct, within the SBURA. Results and conclusions were then drawn, including the potential application of the model to other precincts within the study area. It was suggested that the application of the model to larger areas beyond the Canterbury Renewal Precinct could be applied to: the City of Canterbury Bankstown; the SBURA; and, other Greater Sydney middle ring areas. Finally, results and conclusions on applying the model were summarised.

Addendum 4.1 Canterbury Renewal Precinct and Value Capture Planning (VCP) Tables

The Addendum incorporates three planning principles (and components) from the Value Capture Planning (VCP) model (Chap. 2). These planning principles (within tables) will be applied to the Canterbury Renewal Precinct. The tables enable an authority (i.e. State and municipal) to calculate needs stemming from new housing developments. The three VCP model planning principles as introduced in Chap. 2 are applied, including: *Equitable Housing*, *Public and Open Spaces*, and *Sustainable Transport*. The second stage of the model, Developer Provisions and/or Levy Alternative is the applied. Each of these is now examined in detail.

Equitable Housing

The first planning principle (Equitable Housing) in Table 4.2 incorporates four steps for the planning body to follow in application to Canterbury Renewal Precinct.

Application

A checklist is provided to identify housing types (Step 1). Step 2 is meeting criteria of need for housing types. Step 3 is addressing design guidelines for housing types. The projected units of new housing by types (Step 4) can be calculated by number and %. The planning body can then use these numbers within developer agreements and/or levy alternatives (Table 4.5).

Public and Open Spaces

The second planning principle (Public and Open Spaces) (Table 4.3) incorporates five steps for the planning body to follow in application to Canterbury Renewal Precinct.

Application

A checklist is provided to identify public and open spaces types (Step 1). Step 2 is meeting criteria of need for public and open spaces provision. Step 3 is projecting new public and open spaces by type. Step 4 is addressing design guidelines for public and open spaces. Step 5 calculates the cost of the new public and open spaces. The planning body can they use these numbers within developer agreements and/or levy alternatives (Table 4.5).

Sustainable Transport

The third planning principle (Sustainable Transport) (Table 4.4) incorporates five steps for the planning body to follow in application to Canterbury Renewal Precinct.

Application

A checklist is provided to identify sustainable transport provisions (Step 1). Step 2 is meeting criteria for sustainable transport provisions. Step 3 is projecting sustainable transport provisions for new public by type. Step 4 is addressing design guidelines for sustainable transport provisions. Step 5 calculates the cost of new transport infrastructure. The planning body can they use these numbers within developer agreements and/or levy alternatives (Table 4.5).

Developer Provision and/or Levy Alternative

The forth planning principle (Developer Provision and/or Levy Alternative for Public and Open Spaces) (Table 4.5) incorporates two steps for the planning body to follow in application to Canterbury Renewal Precinct.

Application

A checklist is provided to carry forward the projected units (Step 1) of new housing by type and % from Table 4.2. Step 2 identifies the developer provision or levy alternative per new housing unit for public and open spaces (Table 4.3) and sustainable transport provisions (Table 4.4). The planning body can then use these numbers within developer agreements and/or levy alternative.

References

Canterbury Bankstown City (2017) Draft Sydenham to Bankstown urban renewal corridor strategy. Canterbury, NSW Australia

Department of Planning, Industry and Environment (2014) A plan for growing Sydney. NSW Government, Sydney

Part IV
Regional Growth Centre Development and Value Capture Planning

Part IV examines regional growth centres development and value capture planning (VCP). The VCP approach (outlined in Chap. 1) will apply a model (developed in Chap. 2) to a regional growth centre area. This area is the Central Coast's Somersby to Erina Corridor (under the *Somersby to Erina Corridor Strategy*, SECS). The Gosford City Centre Revitalisation (GCCR) program is also outlined. Finally, the VCP model is applied to the case study area of the Gosford City Centre Precinct (within the GCCR program).

Chapter 5
Regional Growth Centre Renewal and Value Capture Planning – Gosford City Centre Revitalisation (GCCR)

Abstract The chapter addresses revitalisation of regional growth centres using a value capture planning (VCP) model. Two subjects (from earlier chapters) of significance to growth centres are firstly summarised: (1) the implications of the coronavirus pandemic (2020) for future urban planning (Chap. 1); and, (2) VCP model (including components of housing, public and open spaces, and sustainable transport) (Chap. 2). The Gosford City Centre Revitalisation (GCCR) program (under the State of New South Wales and within the Central Coast Council) is then introduced. The Gosford City Centre is then selected as the case study precinct within the GCCR. The position of the City Centre within the Somersby to Erina Corridor (SEC) is then outlined. The VCP model is then applied to the Gosford City Centre precinct. Results and conclusions are then drawn, including the potential application of the model to other precincts within the study area. The application of the model beyond the Gosford City Centre is also addressed, including application to: the SEC; the Central Coast Council area overall; and, other NSW regional growth centres. At chapter's end it's noted that Chap. 7 will address: (1) overall results of applying the model to each of the four case study areas; and, (2) the application of the model to any city or regional growth centre in Australia or in the world.

Dedication There are many people in the history of cities and renewal who have left major legacies. This chapter is dedicated to two of them (Plate 5.1): Brian McGowan (politician) and Phyllis Bennett (teacher and arts advocate)

Brian McGowan (1935–1994) was an Australian politician who was elected as a member of the New South Wales Legislative Assembly. He held this seat until 1988. While in Parliament he was concerned with social and educational issues. A bridge linking Gosford to West Gosford was named in honour of Brian McGowan on 10 December 1995 and carries the Central Coast Highway. (Source: Wikipedia 2020)

Phyllis Albina Bennett (1906- about late 1980s) was a teacher and arts advocate. She was an Arts Council tutor, held classes on English Literature and the creative arts at community college, and served for 17 years on the committee working for the Gosford City Art Centre to be established at Caroline Bay. She was instrumental in forming the Central Coast Arts Council, and founded the Gosford Players in 1970. In 1978 Phyllis was made a Member of the British Empire, in recognition of her services to the arts. (Source: Wikipedia 2020)

R. Rauscher, *Renewing Cities with Value Capture Planning*,
https://doi.org/10.1007/978-3-030-62958-8_5

Plate 5.1 Dedications: Brian McGowan and Phyllis Bennett

5.1 Regional Growth Centres and Value Capture Planning (VCP)

Most cities around the world experience outward movement of their populations into new regional growth centres. The planning of these centres is normally dependent on the state and regional authorities as well as central city administrations. There are thus several regional growth centres outside the Greater Sydney Region (Chap. 2), including the Wollongong-Illawarra and Central Coast regions. As noted in Chap. 2, in summarizing urban planning implications of the coronavirus pandemic (2020) (Chap. 1), planning of Greater Sydney and a growth centre such as the Central Coast (including Gosford City Centre) would need to be reviewed. One implication noted in the chapter, for example, is the likely review of government and private company development (including financing of public and open spaces and transport infrastructure). The role of value capture planning (VCP) could become central to renewing this area (in post coronavirus time) (Plate 5.2).

The Central Coast starts at the Hawkesbury River, 50 k north of the Sydney GPO. Gosford centre (60 km from Sydney GPO) and Wyong Centre 80 km from the GPO. The Wyong centre is 60 km to Newcastle, regional capital of the Hunter Region and (addressed in Chap. 6).

The local planning of this region comes under the Central Coast Council www.centralcoastcouncil.nsw.gov. The Council area (sq. mi) (sq. km) is made up of two districts – Wyong (north) and Gosford (south). The Council was formed as a result of State Government's amalgamation program (2015). Recent State strategic planning documents refer to Gosford as the 'City of Gosford' and also refers to the CBD as 'Gosford City Centre'. For the purposes of assessing urban planning within this State context the author will follow the State's terminology in addressing the Gosford City Centre Precinct. The Central Coast's population in 2020 was 270,000 (2020) and is expected to grow to 330,000 by 2036.

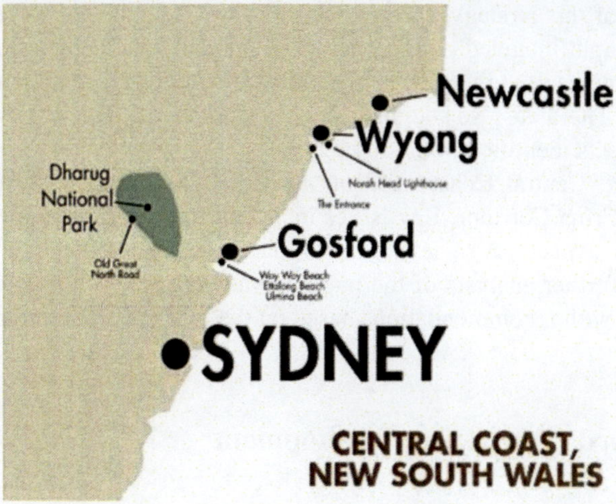

Plate 5.2 Sydney to Gosford Map. (Source: State NSW 2020)

The *Central Coast Regional Plan 2036* (DPIE 2017a) represents a whole of Government commitment to the vision and directions for the Central Coast. It provides an overarching framework that guides the preparation of detailed land use plans; the determination of development proposals; and, informs infrastructure funding. The Regional Plan sets planning priorities for the Central Coast, with medium and longer-term actions to be determined according to rates of growth and change in population and economy over the life of the document.

The Regional Plan is accompanied by a two-year Implementation Plan that sets out governance, tasks, responsibilities and timing for delivery of the Regional Plan. The *Central Coast Implementation Plan 2018–2020* (DPIE 2017b) is the second two-year implementation plan to ensure accountability to realise the vision of the Regional Plan. The priorities have been adjusted towards the next phase of planning for the region's growth and change to 2036.

Achieving the vision for the Central Coast is a shared responsibility between the NSW Government, Central Coast Council, along with business, industry, institutions and the community. Responsibility for delivering the goals are set out in the Regional Plan. This includes 110 goals under seven areas, as follows: economic; land use; housing and employment lands; Aboriginal community needs; city centre revitalisation; works program; and conservation planning.

5.1.1 Somersby to Erina Corridor Strategy (SECS)

The Central Coast Council exhibited the draft Somersby to Erina Corridor Strategy (2019a). This strategy is to deliver the first cohesive long term vision for land use around and along the 25 km stretch of Central Coast Highway. As part of the development of this strategy, community input was sought. The engagement process for this input included online interactive mapping and submissions. Planning factors affecting each of the precincts within the corridor are commented on in the draft strategy. The State also sought advice (in workshops) on how best to prioritise potential projects identified in the strategy.

As a result, Central Coast residents provided feedback on the future of the Somersby to Erina Corridor. Key issues included: transport; parking; active transport; and, green spaces. A review of this Strategy was undertaken in 2020. Council will conduct further analysis of the public comments received and will provide a further report with recommendations based on the feedback received.

5.2 Gosford City Centre Development

Revitalisation of Gosford City Centre is envisaged as the regional capital of the Central Coast. This is a priority of the *Central Coast Regional Plan 2036* and a priority of the Minister for Planning. On 10 October 2018 the Minister for Planning announced the release of the final Government Architect report and approval of a suite of new planning controls to revitalise Gosford City Centre, following 12 months of public consultation on the revitalisation and draft planning measures. The Department of Planning, Industry and Environment (DPIE) (2018a) adopted the Gosford City Centre Revitalisation Program. The program aims to facilitate investment in the city to attract new residents, businesses, tourists and cultural activity. This will allow Gosford to fulfil its potential as the capital of the Central Coast.

Examining further, the State established new planning, development control and assessment arrangements within Gosford City Centre. The controls implement the Government Architect's framework and will (the State says) ensure design led, streamlined, flexible with efficient development parameters for the City Centre. The Department has prepared a report summarising the key comments and issues raised during the 12-month consultation on the Gosford City Centre revitalisation program.

View the Gosford City Centre Revitalisation Community Consultation Report 2018 (PDF, 1.2 MB).

Surveying the development of the Gosford City Centre is an essential first step in considering value capture planning (VCP). Recently a map (Plate 5.3) of development proposals in 2017 showed that the Gosford skyline was set to change ($1b worth of development had been announced). The map shows a sample of the approved developments (those valued over $1.5 m) in the Gosford City Centre South section of Gosford City Centre.

A sample of these development proposals (many had not commenced as of 2020) are now examined. These projects come under development headings (as nominated by the author): (1) Waterfront; (2) Commercial Business District (CBD); (3) New Open Spaces; (4) Heritage Precincts; and, (5) Infrastructure and Services.

Plate 5.3 Map showing approved developments in Gosford City South. (Source: Coast News, Gosford 4 May 2017)

Plate 5.4 Artists' impressions of Gosford development proposals. *Left to right:* a waterfront precinct; and, waterfront hotel. (Source: Central Coast Community News 3 April 2018)

5.2.1 Waterfront

There are a considerable number of proposed projects in the vicinity of the Gosford waterfront (Brisbane Water) with a sample shown in Plate 5.4. An artists' impression shows development of a waterfront precinct (left); and, a waterfront hotel within the St Hillier's site (right).

Plate 5.5 Gosford CBD development proposals. *Left to right*: Gateway Centre; Waterside development earmarked for the old Frogy's site; and, Lederer's Gosford Alive project on the old Kibbleplex site. (Source: Central Coast Council 2020)

5.2.2 Commercial Business District (CBD)

Repositioning (through developments) of the Gosford CBD as a destination for new residents and tourists is illustrated by example in Plate 5.5. The CBD development proposals include (*left to right*): Gateway Centre (left); Waterside development (earmarked for the old Frogy's site) (centre); and, Lederer's Gosford Alive project on the old Kibbleplex site (right) (more comment to follow below on this site, given its size).

The major development of the Kibbleplex site (project cost at $345 m in 2019) is illustrated in Plate 5.6. This includes (left to right): public space (left); buildings' layout proposal (centre); and, proposed walking boulevard (right). Open space planning is examined next.

5.2.3 New Open Spaces

New open spaces are critical in meeting sustainable urban planning principles. The NSW government has published (for example) the final designs for the transformation of part of the waterfront space in Gosford's CBD. The proposal (underway in 2020) is for a new public park (part of the revitalization of Gosford City Centre) (Plate 5.7). Shown here in artists' impressions are of the open space and proposed developments. These include (left to right): new public recreation lands near Leagues Club (left); future development plan for Leagues Club (centre); and, proposed Quarter Masterplan development (right).

Open space planning also provides an underlying foundation for development in the Gosford City Centre. Artists' sketches within the city centre revitalisation area (as noted above) are shown in Plate 5.8. This includes (left to right): water feature (left); grass lands (centre); and further details of the children's aqua feature (right) (as noted above).

The design of these new open spaces are laid out in the Urban Design and Implementation Framework Gosford (NSW Government Architect) (2018) of the

Plate 5.6 Kibbleplex site hub plans. *Left to right:* public space; buildings' layout proposal; and, walking boulevard. (Sources: Infrastructure Magazine; NSW Govt. Architect; and, Central Coast Community News 2020)

Plate 5.7 Artists' impressions of developments creating new open spaces. *Left to right:* new public recreation lands near Leagues Club; future development plan for Leagues Club; and, proposed Quarter Masterplan development. (Source: Central Coast Council 2020)

GCRA plan. This framework covers the rejuvenation of the town's public spaces. This is being done with State Government in partnership with urban designers. The project (using part of the Gosford Leagues Club field) will connect the CBD to Brisbane Water. A key feature of the design is a "tidal terrace" – a shallow water-play area featuring a tidal mechanism that will empty the area of water twice a day. The design of the terrace is influenced by local Indigenous culture, in particular the narrative.

5.2.4 *Heritage Precincts*

Heritage planning is critical for the conservation of a major portion of Gosford City Centre. Central Coast Council has commissioned the development of a Draft Gosford CBD Heritage Interpretation Strategy (2019b). This strategy is to provide a framework and recommendations on bringing Gosford heritage to life. One of several heritage street corners of Gosford is shown here (Plate 5.9).

Plate 5.8 Transformation of Waterfront. *Left to right:* water feature; grass lands; and, further details of the children's aqua feature. (Sources: Gosford City Centre Revitalisation Fact Sheet 2019 (Dept., Plan.), NSW Govt. Arch. Plans, and Central Coast Council 2020)

Plate 5.9 Draft Gosford CBD Heritage Interpretation Strategy illustration. (Source: Central Coast Council 2020)

5.2.5 Infrastructure and Services

All development in Gosford City Centre is dependent on infrastructure and service provisions. Examples of infrastructure projects and proposals are shown in Plate 5.10. These include: Railway Station Upgrade being planned by the Dept. of Planning, Industry and Environment (DPIE) (left); High-Speed Rail Plan to Sydney and Canberra (long term concept of potential route and times) (centre); and, Gosford Hospital redevelopment within the health and wellbeing precinct (a State significant site) (right).

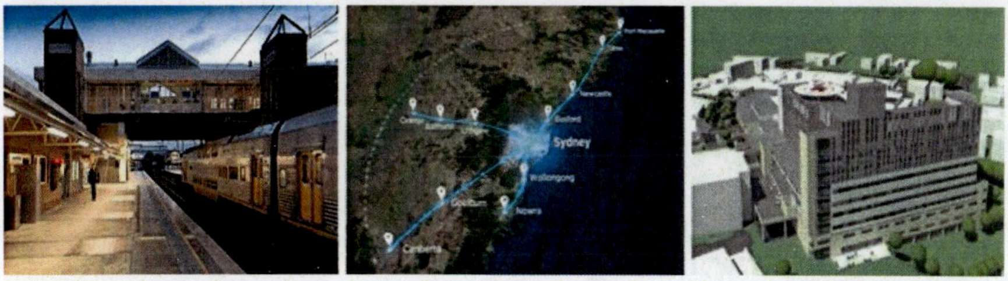

Plate 5.10 Gosford related infrastructure projects and proposals. *Left to right:* Railway Station; High-Speed Rail Plan; and, Gosford Hospital redevelopment precinct. (Source: Central Coast Council 2020)

Plate 5.11 Illustrations of likely high rise in Gosford. *Left to right*: Kibbleplex development (left); 277 Mann St; and, Archibald hotel and residential proposal (Mann and Donnison St.). (Source: Coast News 2020)

Taking three projects (valued at more than $1b) illustrates the likely Gosford skyline transformation as shown in Plate 5.11. This artist's impression (left) firstly shows the Lederer Group's proposed Gosford Alive project (noted above). The proposal covers 5.8 ha in the heart of Gosford. This would see the Kibbleplex building replaced, as noted earlier. The replacement would be apartment blocks housing 730 units in a mixture of high-rise and low-rise buildings. Moving further along, concepts have been released for: (1) developments at 277 Mann St CBD (centre); and, (2) a residential hotel (right) (central to the CBD). All the proposals would significantly change the Gosford Centre's skyline.

Finally, with this transformation to come, the likely increase in the 'metropolitan feel' in the City Centre is forecast by the State. Several additional Gosford high rise projects the State believes will add to this revitalised downtown are noted in Plate 5.12. Moving left to right are: New Hong Kong Macau (mixed use retail, commercial, restaurant, residential development) (left) a residential high rise (centre); and, the completed (2019) Bonython Tower (mixed use development including restaurant, offices and shop top housing) (right).

Plate 5.12 Gosford CBD proposed and built high rise developments. *Left to right*: New Hong Kong Macau development; residential high rise; and, Bonython Tower. (Source: Central Coast Council 2020)

5.3 Gosford City Centre Revitalisation (GCCR) Area Case Study

The State asked for submissions to be made on the State Government's role in revitalisation.

In response, the Central Coast Council (responsible for local governance of the Gosford City Centre) and many community groups within the City Centre have questioned the State Government's planning for this centre. The State Government via the Dept. of Planning, Industry and Environment (DPIE) produced a consultation policy paper (Gosford City Centre Revitalisation Proposed Delivery Mechanisms Main) (2018b) www.planning.nsw.gov.au/gogosford Points particularly relative to value capture planning (VCP) have been noted by the author and are herein summarised (also assistance acknowledged from Jackie Pearson) (Gosford office of the Coast Newspaper).

5.3.1 *Facilities*

There have been both support for and objection to the State financial incentives extended to developers to stimulate Gosford City Centre development. There is a proposed reduction in the overall developer contributions within Gosford City Centre from 4% of the value of development to 3% (in 2019). This would however reduce the funds that would be needed to provide facilities and infrastructure (a value capture principle) as a result of development. On the other hand the establishment of the Gosford City Centre's Special Infrastructure Contribution (SIC) would contribute to needed funds for public needs (facilities and infrastructure). The State notes that the SIC would be applied to projects worth over $1 m. Developers are expected to pay between $20,000 and $1 m in contributions. SIC monies cannot be used for water or sewer, and the consultation paper outlines how SIC payments will be calculated. The Government Architect will decide the final infrastructure list to coincide with the introduction of the SIC. Developers will also be able to 'pay' their SIC in the form of works in kind via agreement. All SIC contributions will be payable before developers receive a Construction Certificate.

5.3.2 *Transport*

Moving to transport, there continues to be a need for an integrated transport planning approach. This is needed in the areas of: car parking; improved bus services (including rapid bus routes connecting centres); water based transport; and, active transport (i.e. cycling and walking). The funding of all these modes of transport requires major State investment and a value capture (VC) mechanism to raise funds from developments (especially within the planning of centres on the Central Coast, including Gosford City Centre). The focus of public transport remains centered on rail hubs. There is scope to upgrade Gosford station, from platforms to a new bus-rail interchange. It's noted that the building of the bus interchange (at the Newcastle Interchange at Wickham) (opened in July 2020). This facility was a developer contribution for that site's residential development bonuses Thus there is scope for a redevelopment under a State-private partnership (via value capture planning) for Gosford Interchange. This would also improve connectivity between the city centre, the Gosford Hospital, and the station (especially if a new Gosford Interchange were undertaken). With needed road upgrades (under the State and local council) there is also critical need for further extensions of pedestrian paths, bicycle lanes and overall improved facilities for disabled access.

5.3.3 *Open Spaces*

Looking at the open spaces in Gosford City Centre there is in need of a major upgrade. Areas such as Gosford City Centre North are particularly short on public domain facilities (i.e. open spaces and play areas). In addition, there are significant opportunities for improvements to link pedestrian access to the Brisbane Water foreshores and across the railway line to the hospital precinct and other facilities. These ideas (especially seeking developer participation) are further explored in the Department of Planning, Industry and Environment (DPIE) (2018c) consultation paper on Gosford City Centre. In addition, long range open space upgrades are noted in the Central Coast Regional Plan 2036 (DPIE 2017a). Finally, the role of the Hunter and Central Coast Regional Development Corporation (HCCRDC) and the State's Delivery, Coordination and Monitoring Committee are important bodies to ensure designed improvements in Gosford City Centre come about. This should include better use of government underutilised lands and capturing value in all land development rezonings and major State infrastructure projects (i.e. transit related bus-rail interchange noted above).

In terms of development heights and floor space allocations (including development bonuses and other value adding incentives) the State (via the Planning Dept.) has adopted (2018) a 'no maximum height or floor space ratio (FSR)'. Under this policy developers are guided as follows:

1. To increase their floor space ratio from 2:1 up to 4:1 for a wholly non-residential building. Medium sites (2800–5600 square metres or 36 metre or greater street frontage) are able to seek height variations subject to a design review panel, additional commercial floor space and environmental measure
2. Large sites (5600 square metres or greater) can access height and floor space variations subject to a design review panel, design competition process, minimum environmental performance standards, additional commercial floor space and a site masterplan.
3. Developments valued at more than \$75 m will be State Significant, assessed by NSW Planning, and determined by the Minister, a delegate or an Independent Planning Commission, if Council, or more than 20 submitters, object, or the applicant discloses a political donation.
4. Those developments worth between \$10 m and \$75 m will be assessed by the Planning Department and determined by the Minister, delegate or Independent Planning Commission if valued at more than \$40 m and Council objects.
5. Those developments valued at less than \$10 m will be assessed and determined by Council. The Joint Regional Planning Panel (JRPP) may still have a role to play in determining some CBD projects in the \$5 million to \$10 million range depending on the assessment process.
6. A Special Infrastructure Contribution (SIC) will be introduced, it will be a levy paid by developers to share the cost of delivering infrastructure. Development contributions collected in Gosford City Centre will be leveraged by the NSW Government, working with Council, to update the Civic Improvement Plan 2007, and introduce a SIC for the Gosford City Centre.
7. The State notes that since 2007, Gosford Council had collected \$8.1 m from developers earmarked for a waterfront precinct and cultural centre. Currently Section 94 contributions are set at 4% but the new SIC will be 2% and an updated local contributions plan will charge an additional 1%.

Moving to financing urban upgrading, the Department of Planning, Industry and Environment (DPIE) (in its policy development) has argued for better use of money collected through developer contributions. The State has also argued for aligning public funding towards infrastructure priorities and, to support in-kind contributions. Listing specific projects and funding mechanisms the State notes:

> Civic improvement works will be accelerated, including: Kibble Park; Leagues Club Field; Burns Park; a potential new public space in the northern precinct; a new pedestrian and cycling bridge connecting the hospital and high schools with the core and north of the CBD; and, upgrades to the Donnison St and Etna St bridges over the railway line. Developer contributions, however, won't be enough for civic works. Work will continue to prioritise works, with funding from the NSW Government, Central Coast Council and developer contributions.

Moving to housing, there has been significant support for more affordable housing to be included in the Strategic Infrastructure Contribution (SIC) (and for more affordable housing in general in Gosford). In terms of housing, the NSW Government intends to investigate opportunities to use the Housing Acceleration Fund to resolve infrastructure blockages in the Gosford City Centre. The State notes that this is a 'short term' way to support the delivery of the 13,000 new dwellings to be built in the CBD in the next 20 years. With this background in development, the value capture planning (VCP) model application is examined next.

5.4 Value Capture Planning (VCP) Model Application

The Value Capture Planning (VCP) model (Tables 2.1, 2.2, 2.3, and 2.4 in Chap. 2) could be considered by the State for the Gosford City Centre. That model is now further outlined and applied to the Gosford City Centre Revitalisation Area. In selecting urban planning principles and components (Chap. 2) these can be added to or modified within the State and Council planning approaches. This planning could include, for example: additional local authority and/or State plans; identification of further VCP precincts; and, projection of key potential development sites. The three VCP model planning principles as introduced in Chap. 2 are applied, including: *Equitable Housing*, *Public and Open Spaces*, and *Sustainable Transport*. The second stage of the model, Developer Provisions and/or Levy Alternative, is then applied. Each of these is now examined in detail.

5.4.1 Equitable Housing

The *Housing* component (with elements of affordable, social and market housing) of the value capture model for the Gosford City Centre revitalisation area is illustrated Addendum 5.1 (Table 5.1). Key factors here include: projected units of new housing by type, number and % of all renewal housing: a. affordable at 20%; social at 10%; and, market at 70%.

5.4.2 Public and Open Spaces

The *Public and Open Spaces* component (with elements of natural areas, green links, and public spaces) of the value capture model for the Gosford City Centre revitalisation area is illustrated Addendum 5.1 (Table 5.2). Key factors here include projected open spaces in categories of: natural areas; open spaces; and, public spaces (quantities to be determined by final plans).

Table 5.1 Value Capture Planning (VCP) Table: Equitable Housing – Gosford City Centre Revitalisation (GCCR) Area

Step	1	2	3	4
Checklist	Housing types	Meet criteria of need for housing types (Examples)	Addressing design guidelines for housing types (Examples)	Projected # and % of new housing units (Example)
	All new housing	Housing standards	Variety of designs	13,000 (100%)
√	1. Affordable housing	Special housing groups	Innovative architecture	2600 (20%)
√	2. Social housing	Social support needs	Maintenance efficiencies	1300 (10%)
√	3. Market housing	Housing as income compatible	Range of housing style choice	9100 (70%)

Steps as per columns	
1	Tick the checklist for housing types that apply
2	Adopt criteria for housing provision
3	Refer to design guidelines for housing
4	Projected units of new housing and % by type

Table 5.2 Value Capture Planning (VCP) Table: Public and Open Spaces – Gosford City Centre Revitalisation (GCCR) Area

Step	1	2	3	4	5
Checklist	Public and open space types	Meeting criteria for public and open spaces provision (Examples)	Projected new public and open spaces	Address design guidelines for public and open spaces (Examples)	Cost of public and open spaces provision by type
√	1. Natural areas	Natural area regeneration	ha	Amount of regeneration	$
√	2. Open spaces	Active and passive spaces	ha	Connecting public and open spaces	$
√	3..Public spaces	Need for public and open spaces	ha	Amount of spaces required	$
Totals			ha		X $ (total)

Steps as per columns	
1	Tick the checklist for public and open spaces types that apply
2	Adopt criteria for public and open spaces provision
3	Projected new public and open spaces
4	Refer to design guidelines for public and open spaces
5	Cost of public and open spaces provision by type (total cost shown as X$)

5.4.3 Sustainable Transport

The *Sustainable Transport* component (with elements of rail, bus and active transport) of the value capture model for the Gosford City Centre revitalisation area is illustrated Addendum 5.1 (Table 5.3). Key factors here include projected sustainable transport in categories of rail, bus and active transport (i.e. cycleways, pedestrian paths and share ways).

5.4.4 Developer Provision and/or Levy Alternative

Having applied the value capture model to housing, public and open spaces and transport, the calculation of a value capture developer provision and/or levy can be completed. A Table (5.4 in Addendum 5.1) incorporates the steps for the State and/or local authority to apply under value capture planning (VCP) for calculating that provision and/or levy per new housing unit.

Table 5.3 Value Capture Planning (VCP) Table: Sustainable Transport – Gosford City Centre Revitalisation (GCCR) Area

Step	1	2	3	4	5
Checklist	Sustainable transport provisions	Meeting criteria for sustainable transport provisions (Examples)	Address design guidelines for sustainable transport (Examples)	Projected new transport provisions (i.e. infrastructure)	Cost of new transport provisions by type
√	1. Rail (heavy and light)	Demand and access	New station, transport or interchange	State/Developer	$
√	2. Bus	Demand and access	Bus transit way	State/Developer	$
√	3. Active transport	Cycle ways, Pedestrian paths, and share ways	Cycling and walking links	Council/ Developer	$
Total					Y $

Steps as per columns	
1	Tick the checklist for sustainable transport types that apply
2	Adopt criteria for sustainable transport provisions
3	Refer to design guidelines for sustainable transport provisions
4	Projected new transport provisions
5	Cost of new transport provisions by type (total cost shown as Y$)

Table 5.4 Value Capture Planning (VCP) Table: Developer Provision and/or Levy Alternative for Open Spaces, and Transport Infrastructure – Gosford City Centre Revitalisation (GCCR) Area

Step	1	2	
		Developer provision or Levy alternative per new housing unit	
Checklist	Projected units of new housing by type and % (see Table 5.1)	Public and open spaces X $ (see Table 5.2)	Transport provision Y $ (total) (see Table 5.3)
✓	All new housing (example) 4000 (100%)	**X $ + Y $** ÷ 13,000 Units = Z $ per unit	
✓	1. Affordable housing 2600 (20%)	Z $ per unit for 2600 Units	
✓	2. Social housing 1300 (10%)	Z $ per unit for 1300 Units	
✓	3. Market housing 9100 (70%)	Z $ per unit for 9100 Units	
Steps as per columns			
1	Tick the checklist for housing types that apply and projected new housing by type and %		
2	Value capture levy per new housing unit for public and open spaces, and sustainable transport provisions (Z $ per unit)		

5.5 Extending the VCP Model Beyond the Precinct

The application of the VCP model to areas beyond the Gosford City Centre precinct can also be considered. This would enable the planning authorities (e.g. State and local councils) to consider applications to: the Somersby to Erina Corridor; the Central Coast Council area overall; and, other NSW regional growth centres.

5.6 Results and Conclusions on Applying the Model

In applying the value capture planning (VCP) model (via the four tables) to the Gosford City Centre case study (under the Gosford City Centre Revitalisation program) there were a number of results and conclusions. Examining the first planning principle (*Equitable Housing*) within the model it was shown that value capture could ensure Gosford City Centre meets affordable, social and market housing needs. At the same time it was shown that the public (via the State) could benefit from land value increases in the rezoning and land upgrading processes under the Gosford City Centre Revitalisation program.

Moving to the second planning principle (*Public and Open Spaces*) within the model it was shown how public and open spaces can be provided at an optimum provision basis. The revitalisation area of the City Centre lends itself to opportunities for new connecting public and open spaces. This includes restoring natural areas such as the Brisbane Water foreshores and bush reserves that have been impacted over the years. Finally, the land value increases from Gosford City Centre's development should enable these public and open space provisions to be achieved.

Moving to the third planning principle (*Sustainable Transport*) the upgrading of the Gosford rail station could be a priority for the State, along with associated new residential and commercial businesses. These would be major contributors to land value increases. These increases can be configured to flow partly back to the pubic (via the State ownership of the station's site and curtilage). Significant increases in State asset value and potential land leases and/or sales could create a further revenue stream of public benefit. Part of this stream could be applied to undertaking further transport initiatives (i.e. bus service facilities, transit ways and active transport) in the Gosford City Centre (and beyond).

Finally, moving to financing, it was shown how the developer provision and/or levy alternative per new housing unit could be incorporated within the model. The total cost of public and open spaces was added to the total cost of sustainable transport infrastructure to achieve an overall cost. It was then shown that these costings could be a combination of developer provision and/or levy alternative per new housing unit.

In conclusion, there is a cumulative positive end result in applying the VCP model to the Gosford City Centre revitalisation area (i.e. in costing and meeting public needs). It was shown that the model could be used for one or more of the three planning principles (dependent of circumstances of precinct revitalisation). It was also suggested that the application of the VCP model beyond the Gosford City Centre could be considered. A comparison of these results with the results of applying the model to the three other case study sites (Chaps. 3, 4, and 6) will be addressed in Chap. 7.

5.7 Summary

The chapter addressed revitalisation of regional growth centres using a value capture planning (VCP) model. Two subjects (from earlier chapters) of significance to growth centres were firstly summarised: (1) the implications of the coronavirus pandemic (2020) for future urban planning (Chap. 1); and, (2) VCP model (including components of housing, public and open spaces, and sustainable transport) (Chap. 2). The Gosford City Centre Revitalisation (GCCR) program (under the State of New South Wales and Central Coast Council) was then introduced. It was noted that the Gosford City Centre was selected as the case study precinct within the GCCR program. The position of the City Centre within the Somersby to Erina Corridor (SEC) was then outlined. The VCP model was then applied to the Gosford City Centre precinct. Results and conclusions were then drawn, including the potential application of the model to other precincts within the study area. The extension of the model to areas beyond the Gosford City Centre was also addressed, including application to: the SEC; the Central Coast Council area overall; and, other NSW regional growth centres.

Addendum 5.1 Gosford City Centre Revitalisation (GCCR) Area and Value Capture Planning (VCP) Tables

The Addendum applies (in Tables) three planning principles and components (in brackets) of those principles as follows (Chap. 2): (1) *Equitable Housing* (affordable, social and market). (2) *Public and Open Spaces* (natural areas, open spaces, and public spaces); and, (3) *Sustainable Transport* (rail, bus and active transport). In addition, the second part of the model is a table incorporating *Developer Provisions and/or Levy Alternative*. The tables enable an authority (i.e. State and municipal) to calculate needs stemming from new housing developments or State infrastructure projects. The tables allow calculation of: (1) projected units of new housing; (2) projected public and open spaces and costing; (3) projected transport infrastructure and costing; and, (4) developer provisions and/or levy alternative. The planning principle Equitable Housing is illustrated first.

Equitable Housing

The first planning principle (equitable housing) in Table 5.1 incorporates four steps for the planning body to follow in application to Gosford City Centre.

Application

A checklist is provided to identify housing types (Step 1). Step 2 is meeting criteria of need for housing types. Step 3 is addressing design guidelines for housing types. The projected units of new housing by types (Step 4) can be calculated by number and %. The planning body can then use these numbers within developer agreements or levy alternatives (Table 5.4).

Public and Open Spaces

The second planning principle (public and open spaces) (Table 5.2) incorporates five steps for the planning body to follow in application to Gosford City Centre.

Application

A checklist is provided to identify public and open spaces types (Step 1). Step 2 is meeting criteria of need for public and open spaces provision. Step 3 is projecting new public and open spaces by type. Step 4 is addressing design guidelines for public and open spaces. Step 5 calculates the cost of the new public and open spaces. The planning body can they use these numbers within developer agreements and/or levy alternatives (Table 5.4).

Sustainable Transport

The third planning principle (sustainable transport) (Table 5.3) incorporates five steps for the planning body to follow in application to Gosford City Centre.

Application

A checklist is provided to identify development provision or sustainable transport provisions (Step 1). Step 2 is meeting criteria for sustainable transport provisions. Step 3 is addressing design guidelines for sustainable transport. Step 4 is projecting sustainable transport provisions new public and open spaces by type. Step 5 calculates the cost of new transport provisions. The planning body can they use these numbers within developer agreements and/or levy alternatives (Table 5.4).

Developer Provision or Levy Alternative

The forth planning principle (developer provision and/or levy alternative for open spaces) (Table 5.4) incorporates two steps for the planning body to follow in application to Gosford City Centre.

Application

A checklist is provided to carry forward the projected units (Step 1) of new housing by type and % from Table 5.1. Step 2 identifies the developer provision or levy alternative per new housing unit for public and open spaces (Table 5.2) and sustainable transport (Table 5.3). The planning body can then use these numbers within developer agreements and/or levy alternative levies.

References

Central Coast Council (2019a) Somersby to Erina corridor strategy. Central Coast Council, Gosford

Central Coast Council (2019b) Draft Gosford CBD heritage interpretation strategy. Central Coast Council, Gosford

Department of Planning, Industry and Environment (2017a) Central coast regional plan 2036. NSW Government, Sydney

Department of Planning, Industry and Environment (2017b) Central coast implementation plan 2018–2020. NSW Government, Sydney

Department of Planning, Industry and Environment (2018a) Gosford city centre revitalisation program. NSW Government, Sydney

Department of Planning, Industry and Environment (2018b) Gosford city centre revitalisation proposed delivery mechanisms. NSW Government, Sydney

Department of Planning, Industry and Environment (2018c) Gosford city centre consultation paper. NSW Government, Sydney

NSW Government Architect (2018) Urban design and implementation framework Gosford. NSW State Government, Sydney

NSW Government Architect (2019) Draft Gosford CBD heritage interpretation. NSW State Government, Sydney

Part V
Regional Capital City Renewal and Value Capture Planning

Part V addresses regional capital city renewal and value capture planning (VCP) (outlined in Chap. 1). A VCP model (developed in Chap. 2) is applied to the Greater Newcastle Metropolitan Area (GNMA). The case study within this area is the Newcastle West Renewal Precinct, a precinct within the Newcastle City Centre Renewal (NCCR) program.

Chapter 6
Regional Capital City Renewal and Value Capture Planning – Greater Newcastle Metropolitan Area (GNMA)

Abstract This chapter addresses the renewal of regional capital city centres using a value capture planning (VCP) model. Two subjects (from earlier chapters) of significance to regional capital city centres are firstly summarised: (1) the implications of the coronavirus pandemic (2020) for future urban planning (Chap. 1); and, (2) VCP model, including three planning principles of: equitable housing, public and open spaces, sustainable transport (developed in Chap. 2). In addition; a second stage of the model incorporating developer provisions and/or levy alternatives is outlined. The study area of the Greater Newcastle Metropolitan Area (GNMA) is then introduced. The case study within this area is the Newcastle West Renewal Precinct under the Newcastle City Centre Renewal (NCCR) program. The Wickham area (within the Renewal Precinct) is selected for examining the funding of public facilities and infrastructure via value capture. The VCP model is then applied to the wider Newcastle West Renewal Precinct. Results and conclusions are then drawn, including the potential application of the model to other precincts within the study area. The application of the model beyond the Renewal Precinct is also addressed, including application to: the NCRA; City of Newcastle; and, the GNMA. The ability to apply the VCP model to other NSW State regional capitals is also addressed. At chapter's end it's noted that Chap. 7 will address: (1) overall results of applying the VCP model to each of the four case study areas; and, (2) the application of the model to any urban or regional centre elsewhere in Australia or in the world.

Dedication There are many people in the history of cities and renewal who have left major legacies. This chapter is dedicated to two of them (Plate 6.1), Joyce Cummings (Mayor and civic worker) and Greg Heys (Mayor and sustainable cities)

Joyce Anne Cummings, AM (1923–2003) was a popular Australian politician in the City of Newcastle. She was the Lord Mayor of Newcastle, and Australia's first female Lord Mayor, from 1974 to 1976 and again from 1977 until 1984, when she retired from politics following a severe stroke. A statute was erected to acknowledge her civic work in late 2019 in CBD of Newcastle. (Source: Wikipedia 2020)

Greg Heys (1945–2007) was Mayor of the City of Newcastle. He was also a director on the Honeysuckle Development Corporation, the Chairman of the Hunter Waste Planning and Management Board, as well as serving on many committees. His leadership as Lord Mayor saw the hugely successful and ground breaking Pathways to Sustainability Conference here in 1997, positioned Newcastle to be head of the pack of all other Australian cities in pursuing ecologically sustainable development policies and practices in the late 1990s. (Source: Wikipedia 2020)

R. Rauscher, *Renewing Cities with Value Capture Planning*,
https://doi.org/10.1007/978-3-030-62958-8_6

Plate 6.1 Dedications: Joyce Cummings and Greg Heys

6.1 Regional Capital Cities and Value Capture Planning (VCP)

Regional capital cities are in a position to individually assess their future development in considering adoption of value capture planning (VCP). Newcastle City (Plate 6.2) is located on the coast of the State of New South Wales, north of the Central Coast (Chap. 5) and 117.1 km north of Sydney.

Newcastle City, the regional capital of the Hunter Region, is shown on this aerial view (Plate 6.3). The city lies adjacent to the Pacific Ocean (bottom of photo) and is served by the Newcastle Harbour (right from bottom to top).

In the 1980's the built Newcastle (Plate 6.4) at the time was basically a medium rise city. The plate shows: the rail service (removed in 2015 as part of the introduction of Newcastle CBD trams) (lower left), CBD (right) and Pacific Ocean (top).

One implication from the coronavirus impact is the likely review of government planning overall (Chap. 1). This would include: examining residential densities and settlement patterns; transport; unban infrastructure provisions: and, public and open space needs. The role of value capture planning i (VCP) could thus become an essential and pivotal planning tool in this review. This will include urban development policies to follow in a post coronavirus era. The Greater Newcastle Metropolitan (GNMA) is examined next.

6.2 Greater Newcastle Metropolitan Plan 2036

The Hunter Valley (Plate 6.5) consists of the Lower Hunter (lower in purple) and the Upper Hunter (left in aqua). North of the Hunter (upper right in blue) is Dungog Shire and the Mid Coast. The Lower Hunter local government areas include: City of

Plate 6.2 Newcastle location within New South Wales. (Source: NSW Government 2020)

Plate 6.3 Newcastle City Aerial. (Source: NSW Government 2020)

Newcastle; City of Lake Macquarie; City of Cessnock; City of Maitland; and, Shire of Port Stephens.

The *Hunter Regional Plan 2036* (2017) is a 20 year blueprint for the future of the Hunter Valley https://www.planning.nsw.gov.au/hunter-region. The plan aims to create a leading regional economy in Australia with a metropolitan city at the heart. The plan acknowledges the extent of the natural environment of the region and the potential to create greater housing choice and jobs. Moving from the Regional Plan, The Greater Newcastle Metropolitan Plan 2036 (2018) is a 20 year

Plate 6.4 Newcastle 1980s. (Source: Fifth Estate)

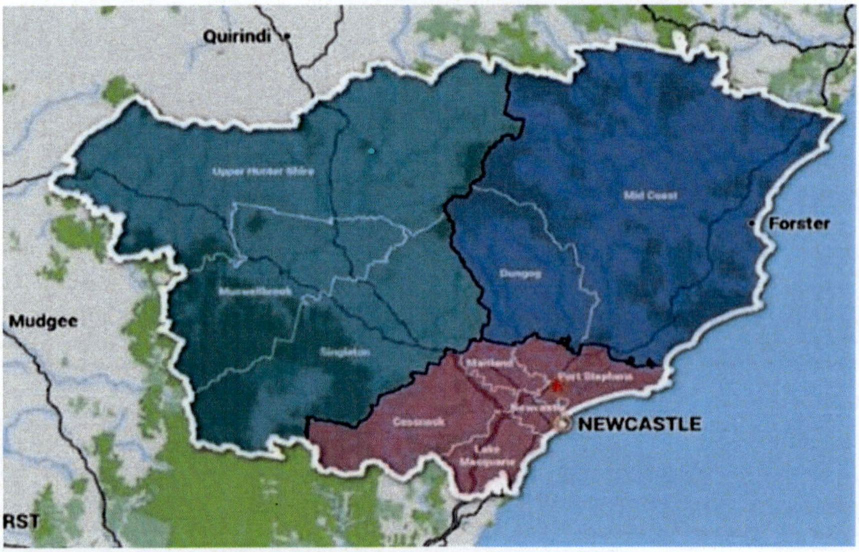

Plate 6.5 Hunter Valley map. (Source: DPIE 2020)

blueprint. This Plan on the future of the Lower Hunter was launched by the State in 2018 and can be viewed at: https://www.planning.nsw.gov.au/Plans-for-your-area/Greater-Newcastle-metropolitan-planning The State promotes Newcastle as Australia's seventh largest city and global gateway for northern NSW. The State vision includes assisting Newcastle to become: a vibrant metropolitan city; with thriving communities; and, greater choice in housing and jobs. The Greater Newcastle Metropolitan Plan sets out strategies and actions aims to drive growth across the five local government areas of the Lower Hunter as a Greater Newcastle.

Within the Plan the State noted Newcastle as Australia's newest and emerging economic and lifestyle city. In addition Newcastle is connected with northern NSW and has an entrepreneurial approach with a globally competitive economy. Aspects of the metropolitan planning for Greater Newcastle are addressed next.

6.2.1 *Metropolitan Planning for Greater Newcastle*

Two research reports were commissioned by the State in considering the prospects of planning within a framework of a Greater Newcastle Metropolitan Area. Firstly, Professor Greg Clark (a global planning adviser) completed the Metropolitan Strategic Planning Case Studies Report for Greater Newcastle NSW (2017) (released by the State). The report makes the case for a metropolitan plan for a Greater Newcastle. This was followed by a report by Marco Dimarsi (2017) on economic parameters of a Greater Newcastle plan. The report (Greater Newcastle Metropolitan Strategy – Economic Prospects to 2036) also reaffirmed the Greater Newcastle concept. Dimarsi argues that creativity and innovation in the services and tourism sectors will assist the future of Australia's leading regional economy. This report was commissioned by the State to assess the draft of the Greater Newcastle Metropolitan Plan 2036 (NSW 2018). The report's key recommendations build on Greater Newcastle's successful transformation from a steel city to a new service economy. This economy is based around the health, education and tourism sectors. The plan can be examined at: https://www.planning.nsw.gov.au/Plans-for-your-area/Greater-Newcastle-metropolitan-planning

The local government areas (noted above) covered by the Greater Newcastle planning area are shown in more detail (purple) in Plate 6.6.

Within this Plan the State promotes Newcastle as Australia's seventh largest city and global gateway for northern NSW. The Plan sets out strategies and actions aimed to drive growth across the five local government areas of the Lower Hunter as a Greater Newcastle. The State notes Newcastle as: (1) dynamic and entrepreneurial, with a globally competitive economy; (2) offering great lifestyles minutes from beaches or bushland, the airport or universities; and, (3) a national leader in the new economy, with smarter cities and carbon neutral initiatives. Newcastle City Centre development is examined next.

6.3 Newcastle City Centre Development

Planning in Newcastle City is under the *Newcastle Local Environmental Plan 2012* (LEP) (2012a). The plan is a legal document that provides rules and standards for uses on private and public land within the Newcastle City Council local government area. The LEP is made up of a written document (instrument) and a series of accompanying maps. In addition, the NSW Government has created planning portal

Plate 6.6 Greater Newcastle planning area of Lower Hunter. (Source: DPIE 2020)

https://www.planningportal.nsw.gov.au/ The portal allows access to information about a specific property, including planning rules, controls and development constraints. There is also an interactive map called a planning viewer which maps the planning controls affecting property. All developments also need to subscribe to the City of Newcastle's *Development Control Plan* DCP (2012b) (and technical manuals). This plan supplements the Newcastle LEP and provides additional information that should be taken into account when preparing a development application. Council notes that a Community Participation Plan (2019) complements the LEP and the DCP.

In terms of Newcastle City Centre development (Plate 6.7) there are three precincts (left to right) West (case study precinct in blue circle), Civic (red circled) and Newcastle East (yellow circled).

The extent of residential development in the Newcastle CBD has been escalating since the early 2000's. This is partly illustrated (Plate 6.8) in a map of development projects (developed or proposed).

One of the major initiatives in investment and development in the Newcastle CBD has been the University of Newcastle (Plate 6.9). Shown here are City Campus Plans at Honeysuckle (in artist sketches) illustrated proposed buildings (left) and public space (right).

The University of Newcastle submitted a development application (March 2020) for its first building of seven buildings. The University site covers part of the former railway line through Honeysuckle. The university acquired a two-hectare super-lot from the government in 2017 to develop this masterplan that will include education facilities, student accommodation and public space. The overall proposal, submitted under the state significant development regime, will double the number of students

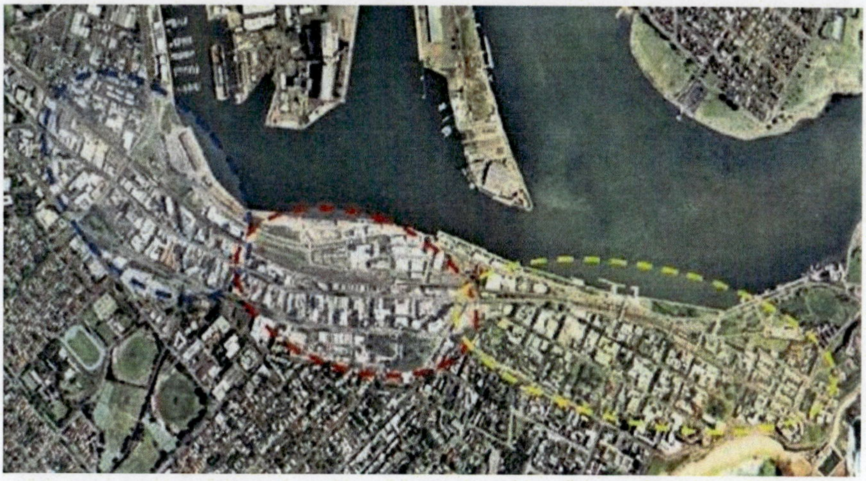

Plate 6.7 Newcastle City Centre three precincts. *Left to right: West,* Civic and East. (Source: Newcastle City Council 2020)

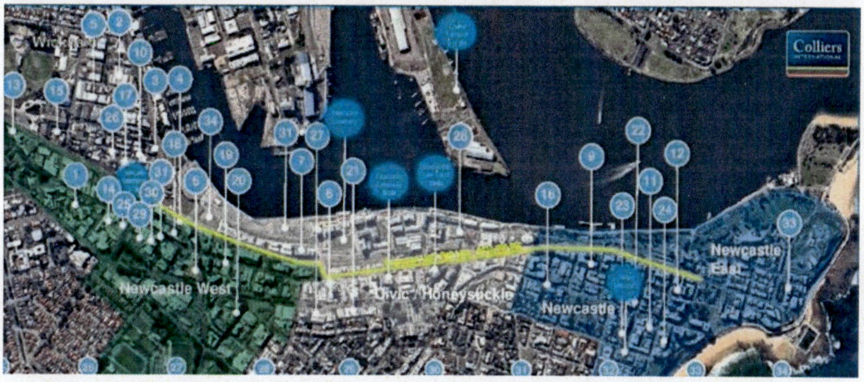

Plate 6.8 Newcastle centre residential (developed or proposed) projects. (Source: Newcastle City Council 2018)

in Newcastle's CBD. The number currently stands at 3000 after the $95 million New Space building opened July 2018.

6.3.1 Newcastle East

A number of development projects in Newcastle are located in Newcastle East (Plate 6.10). Examples of these include (via artists' sketches) (left to right): 1. Old David Jones artist sketch of proposed streetscape; Iris Hotel concept; and, commercial and residential virtual view off Hunter St.

Plate 6.9 Newcastle University's city campus plans at Honeysuckle. *Left to right:* building profile; and, public space. (Source: Newcastle City Council 2020)

Plate 6.10 Newcastle East plans in artists' sketches. *Left to right:* Old David Jones artist sketch of proposed streetscape; Iris Hotel concept; and, commercial and residential virtual view off Hunter St. (Source: Newcastle Council 2020)

The David Jones site (Iris Capital) is a $750 million redevelopment in the vicinity of the Hunter Street Mall. Iris Capital has adopted a master plan for the mainly residential development. The company indicated that the long and narrow block provides the opportunity to create a high-quality streetscape together with residential and commercial outcomes. Iris anticipated that this type of development will help to attract more people into the city centre. That intent supports Council's commitment to the ongoing revitalisation of Newcastle.

Moving further into new developments in the East End (Plate 6.11), these projects will add an additional boost to the CBD. Shown here (left to right) are the Verve residential complex (left); Render House Studios (centre); and, a residential redevelopment (right).

Plate 6.11 New developments (built or proposed) Newcastle East. *Left to right*: Verve Residential; Render House Studios; and, residential east end redevelopment. (Source: Newcastle City Council 2020)

Plate 6.12 Planning development in Newcastle CBD. (Source: Newcastle City Council)

6.3.2 Hunter Street

Moving to Hunter Street, the main road ribbon through the Newcastle CBD, the street continues to attract new development proposals (Plate 6.12). Shown here is a residential complex (left); residential concept plans in Perkins St. (centre); and, pedestrian links to King St. (right). To accommodate the expected new population, transport will be important, as looked at next.

6.3.3 Newcastle Bus and Rail Interchange

The Newcastle bus and rail interchange (Plate 6.13) is considered by the State to be a development catalyst to the Newcastle West and Wickham areas of the city. Shown here are the interchange's main building (left) and entry (right) to the station (with tram connection on the left).

The Newcastle Light Rail is a key part of the Revitalising Newcastle program. The light rail project was delivered by Transport for NSW and services are operated by a private company Newcastle Transport. The light rail commenced in February 2019 providing a frequent travel option into the city centre, connecting key activity

Plate 6.13 Newcastle Interchange at Wickham. (Source: Ray Rauscher 2020)

precincts. The State goal was to reinvigorate Hunter and Scott Streets, and create great urban renewal opportunity.

The former Newcastle Co-Operative Store (known as 'the Store') site has been transformed into the new bus interchange and residential. The State also hopes the interchange will stimulate development. The $200 million redevelopment by the Doma Group is expected to take design to new levels in the city's emerging West End. The project is designed to provide improved connectivity between heavy rail, light rail and buses. In addition, the project was to provide an improved public domain, pedestrian access and upgrading of roads. The group has unveiled plans (Plate 6.14) for Newcastle's largest stand-alone building beside (left) and in what will be the city's tallest apartment towers (right) at Newcastle Interchange.

6.3.4 King Street Vicinity

Moving to King St, development at Newcastle's civic area also accelerated in the 2010's and into the 2020's (Plate 6.15). Planned projects (2020) here include (left to right): conversion of City Hall to a 5 star hotel (left); residential towers (centre): and, a stand-alone residential with commercial below.

6.3.5 Honeysuckle Precinct

Moving to the Honeysuckle Precinct, development started here (and continues in 2020) under the State and Federal Governments' renewal initiatives in the 1980s. This development has continued into the 2020's (Plate 6.16). Here are eight complexes (in concept format) in vicinity of Honeysuckle Drive (includes a cruise terminal proposal at bottom right).

Plate 6.14 Newcastle major developments. (Source: Newcastle Herald 8 Jan 2018)

Plate 6.15 Developments in vicinity of King St. (Source: City of Newcastle 2020)

All of the above projects are a generation apart from Newcastle's earlier steel city image. A glance back at the Newcastle engineering heritage illustrates the extent and dependence that the city had on steel production (Plate 6.17). From left to right is illustrated: ship loading (left); inside working floor (centre); and, exterior surrounds (right).

6.3.6 Affordable Housing

Affordable Housing is major issue in all of the five Greater Newcastle local government areas. This is housing for people on low to moderate incomes such as retail workers, child care and youth workers (all providing essential services). The community housing helps these workers achieve their housing goals, home ownership or private rental. An affordable housing outcome in the heart of the city is a key achievement of the NSW Government's Revitalising Newcastle program. Moving

Plate 6.16 Newcastle City Honeysuckle Dr. apartment concepts and proposed cruise terminal. (Source: City of Newcastle 2020)

Plate 6.17 Newcastle City engineering steel works. (Source: Newcastle Heritage Australia David Moore Steel Works 1963)

Plate 6.18 Newcastle affordable housing examples, Evolve Housing (left frame). (Source: Newcastle City Council 2020)

to an example of this housing, the Hunter Central Coast Development Corporation (HCCDC) works with Evolve Housing, a community housing provider. Evolve was appointed to deliver and manage the development of 30 new affordable rental units in the heart of Newcastle. The Evolve Housing project and one other affordable housing project is shown in Plate 6.18.

Facilitated by the Corporation and delivered by Community Housing Provider, Evolve Housing, the outcome will provide much-needed accommodation for around 80 people. The development of 30 apartments dedicated to affordable rental accommodation will provide a mix of one, two and three bedroom dwellings. In addition, four Specialist Disability Accommodation dwellings will be located on the ground floor. The quality offering will provide an opportunity space for three retail tenancies. To achieve this outcome, the HCCDC contributed both land and Building Better Cities funding to ensure a viable and standalone project. Further details can be sourced at https://www.hccdc.nsw.gov.au/sites/default/files/2020-02/HCCDC-Factsheet-AffordableHousing-20200217.pdf

6.4 Newcastle City's Renewal Strategy (NCRS)

The Newcastle City's Renewal Strategy (2014) provides a number of opportunities for future development of Newcastle (Table 6.1). Some goals (examples) as relative to value capture planning (VCP) have been selected by the author and noted below.

The Newcastle West Renewal Precinct (NWRP) is one of three precincts in the Newcastle City Centre (as noted earlier). The development of the precinct has been (and continues to be in 2020) intense given the number of high-rise buildings under construction (see 6.3 above). Considerable development activity is occurring at

Table 6.1 Newcastle City Renewal Strategy Goals (2014)

1. Removing the heavy rail between Wickham and Newcastle stations and providing a new transport interchange at Wickham for rail, light rail and buses.
2. $340 million from the proceeds of the Port of Newcastle lease, together with $120 million from the Hunter Investment & Infrastructure Fund and Restart NSW for the revitalisation of the Newcastle CBD and the Wickham transport interchange, including heavy rail truncation and the provision of light rail.
3. Amending the Newcastle Local Environmental Plan 2012 to establish a new planning framework to drive renewal of the city centre.
3. Introducing a Development Control Plan for the city centre, which provides detailed planning controls.
4. The State (via Infrastructure NSW) working with investors to facilitate the master planning and redevelopment of key sites (i.e. Newcastle East).

Wickham (Plate 6.19) within the Newcastle West Renewal Precinct. Note the planned intensity of residential high rise along the railway line (left) and Wickham Park (top right). Finally, the low rise urban village (one and two story residential, services, school and local shops) in the centre left is considerably smaller than the village and the 1980s before renewal commenced.

Several of the Wickham planned housing projects are illustrated in Plate 6.20.

A report by AEC (2016) (addressed in Chap. 1) to Newcastle City Council addressed the application of value capture mechanisms within Council's (2018) Wickham Master Plan.

https://www.newcastle.nsw.gov.au/getattachment/Development/Land-Use-Planning/Wickham-Master-Plan-Project/Wickham-Masterplan-Economic-Analysis-Final.pdf.aspx?lang=en-AU

The AEC noted that the Wickham Masterplan will act as the guiding planning document for future redevelopment within Wickham. AEC also notes that the Masterplan will support the principles outlined within the Newcastle Urban Transformation and Transport Program under the Audit Office of NSW (2016).

https://www.audit.nsw.gov.au/our-work/reports/newcastle-urban-transformation-and-transport-program

In examining the implications for funding for public benefits in the Wickham renewal via value capture AEC makes a number of points. These would be applicable to other renewal areas in any locality, as follows:

1. There is an opportunity for Council to capture some of the value created (land and development) to put towards public benefit. This could include specific items of infrastructure that relate to the Wickham Masterplan area.
2. Market attitudes to higher density living continue to shift in favour of city living (i.e. access and convenience). Recent market activity indicates sale prices of apartments in some cases approach or exceed the price of freestanding dwellings on the fringe of the (Newcastle) city centre.
3. There is a recognition by many councils that not all local and community infrastructure can be funded from s94 contributions (Section 7.11 of EPA Act 1979)

Plate 6.19 Wickham Precinct model plan. (Source: Newcastle City Council 2020)

Plate 6.20 Wickham planned housing projects. (Source: Newcastle City Council 2020)

and s94A levies (Section 7.12 of Act). This is particularly relevant for urban renewal areas that historically accommodated industrial and similar uses.

4. The effectiveness of incentive-based mechanisms (i.e. value capture) to fund infrastructure is not in dispute (thus a useful tool).
5. Subject to the environmental capacity of a site, access to additional development potential could be required to contribute to public benefit.

AEC then addresses the effectiveness of an incentive contributions mechanism (i.e. value capture) noting:

1. A clear understanding of the principles and methodology of an incentive contribution system is important for industry confidence and buy-in.
2. The methodology for required contribution should be predictable and transparent, so as to provide certainty to developers and landowners as they contemplate development of their sites.
3. It is important that developers are rewarded for additional risk, particularly in a market such as Newcastle. The higher density residential market in Newcastle is less established and features less depth compared to metropolitan Sydney markets.
4. A possible incentive within the contributions requirement would be an encouragement for contributions in the form of works-in kind (tied into Council's strategic planning). This has a cash flow benefit for developers. It can also serve to deliver key infrastructure works that would benefit the amenity and desirability of the Wickham area.
5. The desirability of residential product is influenced by perception of land use conflicts. Accordingly, the careful planning and management of co-existence by different uses is important. Depending on Council's renewal objectives for the Precinct, a base FSR of 1.5:1 to 2:1 could be considered.
6. It is important to balance the benefits associated with redevelopment and renewal against large scale displacement of existing businesses that may be viable or be part of an economic cluster of similar businesses in the area.
7. Requiring a contribution that is expressed in dollar terms per additional Gross Floor Area (GFA) achieved is considered to be the most transparent and easily understood by the market.
8. When development sites are traded, they are almost without exception quoted in terms of price per unit/site or price per GFA.
9. The mechanism (value capture) would facilitate independent verification by a land economist and/or valuer relying on market evidence.
10. A rezoning or up zoning will almost always immediately translate to increased landowner expectations. This then influences the price paid by a developer (to assemble the development site).
11. It is critical for any intention by Council to 'capture' a proportion of value to be made clear from the outset.

6.5 Value Capture Planning (VCP) Model Application

The Value Capture Planning (VCP) model (Tables 2.1, 2.2, 2.3, and 2.4 in Chap. 2) is applicable to areas such as the Greater Newcastle Metropolitan Area (GNMA). That model (as adopted in Chap. 2) is applied to the Newcastle West Renewal Precinct. The precinct is within the Newcastle City Centre Renewal Plan (2017) as reviewed earlier. The relevant planning includes, for example: the local authority and/or state's current plans; identification of value capture planning precincts; and, projection of key potential development sites. The three VCP model planning principles (first stage of model), including: *Equitable Housing*, *Public and Open Spaces*, and *Sustainable Transport*. The second stage of the model, Developer Provisions and/or Levy Alternative, is then applied. Each of these stages is now examined in detail.

6.5.1 Equitable Housing

The *Housing* component (with elements of affordable, social and market housing) of the value capture model for the Newcastle West Precinct is illustrated in Addendum 6.1 (Table 6.2). Key factors here include (for example): projected units of new housing by type, number and % of all renewal housing: a. affordable at 20%; social at 20%; and, market at 60%.

6.5.2 Public and Open Spaces

The *Public and Open Spaces* principle (with components of natural areas, open spaces, and public spaces) of the value capture model for the Newcastle West Renewal Precinct is illustrated in Addendum 6.1 (Table 6.3).

Table 6.2 Value Capture Planning (VCP) Table: Equitable Housing – Newcastle West Renewal Precinct

Step	1	2	3	4
Checklist	Housing types	Meet criteria of need for housing types (Examples)	Addressing design guidelines for housing types (Examples)	Projected # and % of new housing units (Example)
	All new housing	Housing standards	Variety of designs	10,000 (100%)
✓	1. Affordable housing	Special housing groups	Innovative architecture	2000 (20%)
✓	2. Social housing	Social support needs	Maintenance efficiencies	1000 (10%)
✓	3. Market housing	Housing as income compatible	Range of housing style choice	7000 (70%)

Steps as per columns	
1	Tick the checklist for housing types that apply
2	Adopt criteria for housing provision
3	Refer to design guidelines for housing
4	Projected units of new housing and % by type

Table 6.3 Value Capture Planning (VCP) Table: Public and Open Spaces – Newcastle West Renewal Precinct

Step	1	2	3	4	5
Checklist	Public and open space types	Meeting criteria for public and open spaces provision (Examples)	Projected new public and open spaces	Address design guidelines for public and open spaces (Examples)	Cost of public and open spaces provision by type
✓	1. Natural areas	Natural area regeneration	ha	Amount of regeneration	$
✓	2. Open spaces	Active and passive spaces	ha	Connecting public and open spaces	$
✓	3..Public spaces	Need for public and open spaces	ha	Amount of spaces required	$
Totals			ha		X$

Steps as per columns	
1	Tick the checklist for public and open spaces types that apply
2	Adopt criteria for public and open spaces provision
3	Projected new public and open spaces
4	Refer to design guidelines for public and open spaces
5	Cost of public and open spaces provision by type (total cost shown as X$)

6.5.3 *Sustainable Transport*

The *Sustainable Transport* principle (with components of rail, bus and active transport) of the value capture model for the Newcastle West Renewal Precinct is illustrated Addendum 6.1 (Table 6.4).

6.5.4 *Developer Provision and/or Levy Alternative*

Having applied the value capture model to housing, public and open spaces and transport, the consideration of developer provisions and/or levy alternative can be addressed. A Table (6.5 in Addendum 6.1) incorporates the steps for the State and/or local authority to follow up in cooperation with the development industry.

6.6 Extending the VCP Model Beyond the Study Precinct

The application of the VCP model to areas beyond the Newcastle West Renewal Precinct can also be considered. This would enable the planning authorities (e.g. State and local councils) to consider applications to the: Newcastle City Renewal Area (NCRA); City of Newcastle; and, Greater Newcastle Metropolitan Area (GNMA).

Table 6.4 Value Capture Planning (VCP) Table: Sustainable Transport – Newcastle West Renewal Precinct

Step	1	2	3	4	5
Checklist	Sustainable transport provisions	Meeting criteria for sustainable transport provisions (Examples)	Address design guidelines for sustainable transport (Examples)	Projected transport provisions (i.e. infrastructure)	Cost of new transport provisions by type
✓	1. Rail (heavy and light)	Demand and access	New station, transport or interchange	State/Developer	$
✓	2. Bus	Demand and access	Bus transit way	State/Developer	$
✓	3. Active transport	Cycleways, Pedestrian paths, and share ways	Cycling and walking links	Council/ Developer	$
Total					Y $
Steps as per columns					
1	Tick the checklist for sustainable transport types that apply				
2	Adopt criteria for sustainable transport provisions				
3	Refer to design guidelines for sustainable transport provisions				
4	Projected new transport provisions				
5	Cost of new transport provisions by type (total cost shown as Y$)				

Table 6.5 Value Capture Planning (VCP) Table: Developer Provision and/or Levy Alternative for Public and Open Spaces, and Transport Infrastructure – Newcastle West Renewal Precinct

Step	1	2	
		Developer provision or levy alternative per new housing unit	
Check list	Projected units of new housing by type and % (see Table 6.2)	Public and open spaces X $ (see Table 6.3)	Transport infrastructure Z $ (total) (see Table 6.4)
✓	All new housing (example) 10,000 (100%)	**X $ + Y $** ÷ 10,000 units = Z $ per unit	
✓	1. Affordable housing 2000 (20%)	Z $ per unit for 2000 units	
✓	2. Social housing 1000 (10%)	Z $ per unit for 1000 units	
✓	3. Market housing 7000 (70%)	Z $ per unit for 7000 units	
1	Tick the checklist for housing types that apply and projected new housing by type and %		
2	Value capture levy per new housing unit for public and open spaces, and transport infrastructure		

6.7 Results and Conclusions on Applying the VCP Model

In applying the value capture planning (VCP) model (via the three planning principles and developer provisions and/or levy alternative) to the Newcastle West Renewal Precinct case study there were a number of results. Examining the first planning principle (*Equitable Housing*) within the model it was shown that value capture could ensure Newcastle City Centre renewal sites meet housing needs (affordable, social and market housing). At the same time it was shown that the public (via the State) could benefit from land value increases in the rezoning and land upgrading processes under the Newcastle City Centre Renewal Plan.

Moving to the second planning principle (*Public and Open Spaces*) it was shown how public and open spaces provisions could meet designated needs. As a renewal area Newcastle West lends itself to opportunities for new connecting public and open spaces (including restoring natural areas that have been impacted over the years).

Moving to the third planning principle (*Sustainable Transport*) the creation of the Wickham Interchange (train, light rail and bus) facility in 2019 was noted as is a major contributor to land value increases. These increases should be configured to flow partly back to the public purse. This could occur for example given the State ownership of the rail and bus interchange facilities and land curtilage. Significant increases in State asset value and potential land leasing and/or sales could create a significant revenue stream for the public benefit. Part of this stream could be applied to undertaking further transport initiatives (i.e. bus services facilities, transit ways and active transport) in the Newcastle City Centre (and beyond).

Finally, moving to the developer, the model showed how the developer provision and/or levy alternative per new housing unit could be achieved. For the Precinct, the total cost of public and open spaces was added to the total cost of transport infrastructure to achieve an overall costing associated of these two items. To then achieve a spread of the costings per housing unit this total cost is divided by the anticipated number of new units (e.g. affordable, social and market). It was then shown that these final costings could be met via developer provision and/or levy alternative per new housing unit (final column in that Table). A comparison of these results with the results of applying the model to the three other case study sites (Chaps. 3, 4, and 5) will be addressed in Chap. 7.

6.8 Summary

This chapter addressed the renewal of regional capital city centres using a value capture planning (VCP) model. Two subjects (from earlier chapters) of significance to regional capital city centres were firstly summarised: (1) the implications of the coronavirus pandemic (2020) for future urban planning (Chap. 1); and, (2) VCP model (including planning principles of equitable housing, public and open spaces, sustainable transport, and developer provisions and/or levy alternatives) (Chap. 2). The study area of Greater Newcastle Metropolitan Area (GNMA) was then introduced. The case study, Newcastle West Renewal Precinct under the Newcastle City Centre Renewal (NCCR) area, was then selected. The VCP model was then applied to the Precinct. Results and conclusions were then drawn, including the potential application of the model to other precincts within the study area. The application of the model beyond the Newcastle West Renewal Area was also addressed, including application to the: Newcastle City Renewal Area (NCRA); City of Newcastle overall; and, Greater Newcastle Metropolitan Area (GNMA). Finally, results and conclusions on applying the model were summarised.

Addendum 6.1 Newcastle West Renewal Area and Value Capture Planning (VCP) Tables

The Addendum incorporates three planning principles and developer provisions and/or levy alternative (within tables) within the Value Capture Planning (VCP) model (Chap. 2). These components (in tables) will be applied to the Newcastle West Renewal Precinct. The tables enable an authority (i.e. State and municipal) to calculate needs stemming from new housing developments. The authority is able to apply the tables under the three planning principles: 6.1 Equitable Housing; 6.2 Public and Open Spaces; 6.3 Sustainable Transport. In addition the Developer Provision and/or Levy Alternative is applied (6.4). The tables (in above order) allow calculation of: (1) projected units of new housing; (2) projected public and open spaces and costing; (3) projected transport infrastructure and costing; and, (4) developer provisions and/or levy alternative. The planning principle Equitable Housing is illustrated first.

Equitable Housing

The first planning principle (equitable housing) in Table 6.2 incorporates four steps for the planning body to follow in application to Newcastle West Renewal Precinct.

Application

A checklist is provided to identify housing types (Step 1). Step 2 is meeting criteria of need for all housing types. Step 3 is addressing design guidelines for housing types. The projected units of new housing by types (Step 4) can be calculated by number and % of each type. The planning body can then use these numbers within developer agreements and/or levy alternatives (Table 6.5).

Public and Open Spaces

The second planning principle (public and open spaces) (Table 6.3) incorporates five steps for the planning body to follow in application to Newcastle West Renewal Area.

Application

A checklist is provided to identify public and open spaces types (Step 1). Step 2 is meeting criteria of need for public and open spaces provision. Step 3 is projecting new public and open spaces by type. Step 4 is addressing design guidelines for public and open spaces. Step 5 calculates the cost of the new public and open spaces. The planning body can they use these numbers within developer agreements or levy alternatives (Table 6.5).

Sustainable Transport

The third planning principle (sustainable transport) (Table 6.4) incorporates five steps for the planning body to follow in application to Newcastle West Renewal Area.

Application

A checklist is provided to identify sustainable transport provisions (Step 1). Step 2 is meeting criteria for sustainable transport provisions. Step 3 is addressing design guidelines for sustainable transport provision. Step 4 is projecting sustainable transport provisions by type. Step 5 calculates the cost of new transport provisions. The planning body can they use these numbers within developer agreements and/or levy alternatives (Table 6.5).

Developer Provision and/or Levy Alternative

The developer provision and/or levy alternative (public and open spaces and transport infrastructure) (Table 6.5) incorporates two steps for the planning body to follow in application to Newcastle West Renewal Area.

Application

A checklist is provided to carry forward the projected units (Step 1) of new housing by type and % from Table 6.2. Step 2 identifies the developer provision or levy alternative per new housing unit for public and open spaces (Table 6.3) and sustainable transport (Table 6.4). The planning body can then use these numbers within developer agreements and/or levy alternative levies.

References

AEC (2016) Wickham masterplan economic and market analysis. AEC, Newcastle

Audit Office of NSW (2016) Newcastle urban transformation and transport program. NSW Government, Sydney. https://www.audit.nsw.gov.au/our-work/reports/newcastle-urban-transformation-and-transport-program

Clark G (2017) Metropolitan strategic planning case studies report for greater Newcastle NSW. NSW Government, Sydney

Demonski M (2017) Greater Newcastle metropolitan strategy – economic prospects to 2036. NSW Government, Sydney

Department Planning Industry and Environment (2017) Newcastle city centre renewal plan. NSW Government, Sydney

Department Planning Industry and Environment (2018) Greater Newcastle metropolitan plan 2036. NSW Government, Sydney

Newcastle City Council (2012a) Local environment plan (LEP). Newcastle City Council, Newcastle

Newcastle City Council (2012b) Development control plan. Newcastle City Council, Newcastle

Newcastle City Council (2018) Wickham master plan. Newcastle City Council, Newcastle

Newcastle City Council (2019) Community participation plan. Newcastle City Council, Newcastle

Part VI
Cities and Value Capture Planning Directions

Part VI reviews conclusions reached from the application of the Value Capturing Planning (VCP) model to the case studies and future directions of value capture (**Chap.** 7).

Chapter 7
Future Directions of Planning Cities and Regions with Value Capture

Abstract This chapter reviews conclusions reached from the application of the value capture planning (VCP) model to the four case study precincts and addresses the future of planning cities with value capture. Moving from these precincts (Waterloo, Canterbury, Gosford City Centre, and Newcastle West), each individual planning principle of the model as applied is reviewed, including components (in brackets): *Equitable Housing* (affordable, social and market housing); *Open and Public Spaces* (natural areas, public and open spaces); and, *Sustainable Transport* (rail, bus and active transport). In addition, given implications of the coronavirus (COVID-19) in planning urban areas, comments on the use of VCP in future years are offered. An overall assessment of the VCP model is then presented, including the application of the model to any urban centre in Australia or anywhere in the world.

Dedication There are many people in the history of cities and renewal who have left major legacies. This chapter is dedicated to two of them (Plate 7.1), Patrick Geddes (city planner and philanthropist) and Hebert Gans (sociologist and social planner)

Sir Patrick Geddes (1854–1932) was a British biologist, sociologist, geographer, philanthropist and pioneering town planner. He is known for his innovative thinking in the fields of urban planning and sociology. He introduced the concept of "region" to architecture and planning and coined the term "conurbation". Later, he elaborated 'neotechnics' as the way of remaking a world apart from over-commercialization and money dominance. In the 1920s he bought the Château d'Assas to set up a centre for urban studies. (*Source* Wikipedia 2020)

Herbert J. Gans (1927-) is a German-born American sociologist who taught at Columbia University from 1971 to 2007. One of the most prolific and influential sociologists of his generation, Gans came to America in 1940 as a refugee from Nazi Germany and has sometimes described his scholarly work as an immigrant's attempt to understand America. He trained in sociology at the University of Chicago, where he studied with David Riesman and Everett Hughes, among others, and in social planning at the University of Pennsylvania, where he studied primarily with Martin Meyerson. (Source: Wikipedia 2020)

R. Rauscher, *Renewing Cities with Value Capture Planning*,
https://doi.org/10.1007/978-3-030-62958-8_7

Plate 7.1 Dedications: Patrick Geddes and Herbert Gans

7.1 Background to Value Capture Planning (VCP) Model

Having developed a value capture planning (VCP) model (Chap. 2) and applied that model to case study areas (Chaps. 3, 4, 5, and 6) conclusions were reached for each study area. The results taken for all case studies can now be addressed (7.2 below). In review, the four precinct case study areas (city in brackets) will be looked at as follows: Waterloo Housing Estate Renewal Precinct (City of Sydney); Canterbury Renewal Precinct (City of Canterbury Bankstown); Gosford City Centre (under the Central Coast Council); and, Newcastle West Renewal Precinct (City of Newcastle). The three planning principles with the model (3 Tables) and the developer provisions and/or levy alternative (forth Table) can also be reviewed across the four case study areas (7.3). With these reviews comment can also be offered (across the four study areas) on the implications the coronavirus (Covid-19) for future urban planning and value capture (refer to Chap. 1) (7.4). An overall assessment of the VCP model can then be presented, including future use of value capture in urban planning anywhere (7.5). The application of the VCP model to the precinct case study areas is examined first.

7.2 Results of VCP Model Application to Case Study Precincts

There are a number of results stemming from the application of the value capture planning (VCP) model to the case study precincts (as outlined in Chaps. 3, 4, 5, and 6) and summarised here by precinct.

7.2.1 Waterloo Housing Estate Renewal Precinct

Results of the application of the VCP model to the Waterloo Housing Estate Renewal Precinct (Chap. 3) are examined below. The VCP model's three planning principles and developer provision and/or levy alternative are addressed.

7.2.1.1 Equitable Housing

It was shown that value capture could ensure renewal lands meet housing needs in Waterloo Precinct (e.g. affordable, social, and market housing). It was also shown that the public (via the State) could benefit from land value increases in the Waterloo rezoning of lands and land upgrading within the Waterloo renewal precinct.

7.2.1.2 Public and Open Spaces

Acknowledging the planned high densities of this Waterloo renewal precinct the VCP model could ensure generous open spaces and public spaces are achieved. The Waterloo redevelopment site (given its large size) lends itself to multiple opportunities for connecting public and open spaces. This could include replicating natural area features lost over years (see City of Sydney proposals in Chap. 3). Alternative open space design options (i.e. State and the City Council) provide a means of reaching optimum open space outcomes (and tied into the value capture application).

7.2.1.3 Sustainable Transport

The development of the new Waterloo Metro rail station (and associated residential towers and commercial service businesses) will be a major catalyst to land value increases. These increases should flow to the pubic (given the State ownership of most of the renewal site). There is likely to be significant increases in State asset value and land leasing and/or sales within the project. The Sydney Metro City area should create an additional significant revenue stream for the State (thus public gain). Part of this stream could be applied to undertaking wider transport initiatives in the Redfern and Waterloo locality (i.e. bus services and active transport).

7.2.1.4 Developer Provision and/or Levy Alternative

Having applied the three VCP planning principles, the developer provision and/or levy alternative offers the developer in the Waterloo Renewal Precinct a wide choice in collaborating and contributing to meeting needs. These needs (outlined in Chap. 3) include public and open spaces and sustainable transport provisions. Additional

Waterloo Renewal Precinct opportunities in value capture (within and extending beyond the renewal area) could be taken up by the State, local government planning body or developers.

7.2.1.5 Extending the Model to Other Areas

It was also shown that the model could be applied to areas beyond the Waterloo Housing Estate Precinct, including application to: the CERA; the City of Sydney overall; and, other inner city areas of Greater Sydney.

7.2.2 Canterbury Renewal Precinct

Examining the application of the VCP model to the Canterbury Renewal Precinct (Chap. 4) the results are summarized below. Under the VCP model three planning principles and developer provision and/or levy alternative are addressed.

7.2.2.1 Equitable Housing

It was shown that value capture could ensure renewal lands (e.g. within the Canterbury Renewal Precinct) met housing needs (e.g. affordable, social, and market housing). It was also shown that the public (via the State) can benefit from land value increases in the rezoning of lands and land upgrading within the renewal precinct.

7.2.2.2 Public and Open Space

Given the Canterbury Renewal Precinct will cater for a high density population in, it was shown how public and open spaces can be designed to meet anticipated needs. As a renewal precinct in a middle ring city area, the precinct lends itself to opportunities for new connecting public and open spaces. This includes restoring natural areas such as the Cooks River foreshores that have been impacted over the years. The land value increases from the precinct's development should enable these public and open space provisions to be achieved.

7.2.2.3 Sustainable Transport

The development of the upgraded Canterbury Rail Station (under the Sydney Metro Southwest project) and associated new residential and commercial businesses would be major contributors to land value increases in and adjacent to the Canterbury

Renewal Precinct. These value increases could be configured to flow mostly back to the pubic (i.e. via the State ownership of the station's site and land curtilage). Significant increases in State asset value and potential land leases and/or sales around the Metro station could also create a significant revenue stream of public benefit. Part of this stream could be applied to undertaking further transport initiatives (i.e. bus services and active transport) within the Precinct and beyond.

7.2.2.4 Developer Provision and/or Levy Alternative

The developer provision and/or levy alternative offers the developer in the Canterbury Precinct a wide choice in providing for public and open spaces as well as sustainable transport infrastructure. Additional opportunities (within and extending beyond the VCP model) for these type of provisions would exist for the developer, Canterbury Bankstown Council and the State to examine.

7.2.2.5 Extending the Model to Other Areas

It was also shown that the model could be applied to areas beyond the Canterbury Renewal Precinct, including: (1) Canterbury Bankstown City; (2) Sydenham to Bankstown Urban Renewal Corridor (SBURC); and, other middle ring areas of Greater Sydney.

7.2.3 *Gosford City Centre Revitalisation Precinct*

Examining the application of the VCP model to the Gosford City Centre Revitalisation Area (Chap. 5) the results are summarized below (under the VCP model's three planning principles and developer provision and/or levy alternative).

7.2.3.1 Equitable Housing

It was shown that value capture could ensure renewal lands (e.g. within the Gosford City Centre Revitalisation Area) can meet housing needs (e.g. affordable, social, and market housing). It was also shown that the public (via the State) could benefit from land value increases in the rezoning and land upgrading processes of this growth centre revitalisation area.

7.2.3.2 Public and Open Spaces

Given the Gosford City Centre revitalisation area will cater for a higher density population then currently exists in the centre, it was shown how public and open spaces can be designed to meet generated needs. As a revitalisation initiative the area lends itself to opportunities for new connecting public and open spaces. This includes restoring natural areas such as foreshores of Brisbane Water that have been impacted over the years (a current State project on the Leagues Club fields was noted). The land value increases (as captured by the State) from the City Centre's development should enable these public and open space provisions to be achieved.

7.2.3.3 Sustainable Transport

There is potential for an upgraded Gosford rail and bus interchange. This upgrading could utilise aspects of a Sydney Metro style project under Future Transport 2056 (NSW 2018) (Chap. 2) or the Newcastle Bus Interchange as provided by the developer (Chap. 6). Any upgrading usually includes associated new residential and commercial businesses (contributing to land value increases for the Stated). These increases could be configured to flow mostly back to the pubic (i.e. via the State ownership of the station's site and curtilage). Significant increases in State asset value and potential land leases and/or sales around a new Gosford Interchange would create a revenue stream of public benefit. Part of this stream could be applied to undertaking further transport initiatives (i.e. bus services and active transport) in the Gosford revitalisation area and Central Coast generally.

7.2.3.4 Developer Provision and/or Levy Alternative

The developer provision and/or levy alternative offers the developer a wide choice in collaborating and contributing to meeting public and open spaces, and sustainable transport needs in the Gosford City Centre Revitalisation Area. New provision opportunities (within and extending beyond this revitalisation area) exist for the developer, Central Coast Council or the State.

7.2.3.5 Extending the Model to Other Areas

It was also shown that the model could be applied to areas beyond the Gosford City Centre Precinct, including: Somersby to Erina Corridor; the Central Coast Council overall; and, other regional growth centres of NSW.

7.2.4 *Newcastle West Renewal Precinct*

Results of the application of the VCP model to the Newcastle West Renewal Precinct (Chap. 6) are examined below. Under the VCP model three planning principles and developer provision and/or levy alternative are addressed.

7.2.4.1 Equitable Housing

It was shown that value capture can ensure renewal lands (e.g. within the Newcastle West Renewal Precinct) could meet housing needs (e.g. affordable, social, and market housing). It was also shown that the public (via the State) could benefit from land value increases in the rezoning and land upgrading process of this renewal precinct.

7.2.4.2 Public and Open Spaces

Given the Newcastle West Renewal Precinct will cater for a higher density population than currently exists, it was shown how public and open spaces could be designed to meet the new needs. As a renewal precinct the area lends itself to opportunities for new connecting public and open spaces (including restoring public and open spaces areas in the precinct that have been impacted over the years, such as around Cottage Creek, Wickham).

7.2.4.3 Sustainable Transport

The Newcastle Interchange (train, light rail and bus facility) and associated new residential and commercial businesses are major contributors to land value increases in the Newcastle West Renewal Precinct. These increases could be configured to flow mostly back to the pubic (i.e. via the State ownership of the interchange's and rail station's site and curtilage). Significant increases in State asset value and potential land leases and/or sales around the new interchange could create a revenue stream of public benefit. Part of this stream could be applied to undertaking further transport initiatives (i.e. bus services and active transport) in the renewal area and beyond.

7.2.4.4 Developer Provision and/or Levy Alternative

The developer provision and/or levy alternative in the Newcastle West Renewal Precinct offers the developer a wide choice in contributing to meeting public and open spaces and sustainable transport provision needs in the renewal area. New opportunities to meet public needs (within and extending beyond the renewal area) exists for developers, Newcastle City Council and the State.

7.2.4.5 Extending the Model to Other Areas

It was also shown that the model could be applied to areas beyond the Newcastle Renewal Precinct, including: (1) Newcastle Renewal Area; (2) Newcastle City overall; (3) four other local government areas (Lake Macquarie City, Cessnock City, Maitland City and Port Stephens Shire within the Greater Newcastle Metropolitan Area) (GNMA); and, other regional capitals of NSW.

7.3 Results from VCP Application by Planning Principles and Developer Provisions and/or Levy Alternative

Having reviewed the results stemming from the application of the value capture planning (VCP) model to the four case study precincts, the individual planning principles and developer provisions and/or levy alternative can be examined. The planning principles within the model and components (in brackets) included: Equitable *Housing* (affordable, social and market housing); *Open and Public Spaces* (natural areas, open spaces and public spaces); and, *Sustainable Transport* (rail, bus and active transport). These are now examined, commencing with equitable housing.

7.3.1 Equitable Housing

The VCP model applied the planning principle of *Equitable Housing* to the four case study precincts (noted above). In summary, it can be concluded that the value capture model can create outcomes ensuring renewal and revitalisation lands (e.g. within the case study renewal precincts) to meet affordable, social, and market housing needs. It can also be concluded within all study areas that the public (via the State) could benefit from land value increases in the rezoning of lands and land up grading within the study areas. Finally, it is concluded that a value capture program needs to reflect the varied physical, social and economic circumstance of a locality (i.e. renewal area). This is likely to result in equitable housing provisions varying across a city or region (i.e. inner city and regional growth areas).

7.3.2 Public and Open Spaces

The VCP model applied the planning principle of *Public and Open Spaces* to four case study precincts (noted above). In summary, all case study precincts were planned for higher densities than was current in those precincts. This meant that the value capture program would need to ensure that generous open spaces and public spaces are incorporated into the planning and to be financed. The varying circumstances of the renewal areas, means a value capture program needs to accommodate the multiple opportunities for connecting public and open spaces. In several of the case studies (i.e. Canterbury and Gosford) the importance of restoring natural area features lost over years was noted. Alternative public and open space design options (i.e. proposed by State and local government councils) provides an opportunity for value capture to be used as a planning tool to achieve desired outcomes (i.e. capturing land value increases).

7.3.3 Sustainable Transport

The development (or upgrading) of rail stations featured in all case study precincts. The new Waterloo Metro rail station, Canterbury station upgrading to Metro standards, Gosford Interchange (potential), and the Newcastle Interchange were all noted as potential catalysts to land value increases. It was shown that these increases could flow to the pubic (given the State ownership of most of these renewal sites). It was also shown (at these precinct locations) that there was likely to be significant increases in State asset value and land leasing and/or sales stemming from these transport initiatives. It was concluded that part of this revenue stream could be applied to undertaking wider transport initiatives (i.e. bus services and active transport) within those precincts and beyond.

7.3.4 Developer Provision and/or Levy Alternative

The developer provision and/or levy alternative in all precinct case studies offered the developer options in participation with the State and/or local authority. It was acknowledged that collaboration within State and local government units would vary by location. There was however in all the case study precincts opportunities for developers to contribute in meeting public and open spaces, and transport infrastructure needs. Finally, it was noted there would likely be new opportunities for developers participating within value capture planning (VCP) programs.

7.4 VCP and Implications of Coronavirus to Future Urban Planning

The implications of the coronavirus (COVID 19) to future urban planning was addressed in Chap. 1 (and noted in subsequent chapters). The chapter concluded that it was the denser urban sections of cities that felt the brunt of the virus. This was especially the case in poorer sections of cities, where housing overcrowding may have contributed to the virus spread. The COVID 19 virus's impact on centers raised questions about future planning strategies. The earlier chapter noted the importance of planning to incorporate resilience of the city structure. This would be particularly important for the city to cater to the needs of its population when disasters such as an infectious disease strike. The Chapter also noted the importance of planning administrations being aware of infectious diseases and pandemics. In summary the Chapter noted factors for planners to consider in designing and renewing cities to minimize impacts of future infectious diseases. These factors included: health guidelines for renewing cities; infectious disease control programs; implications of global warming; and, housing design. Future directions of value capture planning (VCP) is looked at next.

7.5 Value Capture Planning (VCP) as a Valuable Planning Tool

The results of the application of the Value Capture Planning (VCP) model to the four case study precincts (noted above) illustrated the value of this planning tool for future planning. The application of value capture planning has shown VCP's advantages as a planning tool, including:

1. *Assists Urban Planning Processes*

The value capture planning (VCP) tool assists the urban planning processes in ensuring needs generated from development are met (i.e. capturing land value increases from government planning actions or public infrastructure construction). For example, it was illustrated how value capture could be applied within renewal corridors. This includes corridors of: Central to Eveleigh, Sydenham to Bankstown; and, Somersby to Erina as well the Greater Newcastle Metropolitan Area (City of Newcastle, Lake Macquarie City, Cessnock City, Maitland City and Port Stephens Shire).

2. *Strengthening Planning within a Renewal Area*

The designation of city or regional renewal (or revitalisation) areas can be significantly strengthened when value capture is introduced as a planning tool.

3. *Flexible Agreements with Developers*

The results of the application of the VCP model illustrated how value capture could be valued by developers and used within development agreements.

4. *Relevant to Local Geography and Development Patterns*

The value capture planning principles (as contained within the VCP model) could be selected as relevant to the area's geography and urban development patterns of the subject area.

5. *Extending the Model to Other Areas*

The VCP model could be applied to areas beyond a study area, as illustrated in applications to the four case study precincts.

6. *Easy to Apply*

Value capture tables (four in this VCP model) had a wide application range and proved relatively easy to apply (in steps).

7. *Additional Planning Principles*

A range of additional planning principles could be added to a value capture planning (VCP) program. The VCP model tables allow the introduction of these additional principles.

8. *Value Capture Planning at Different Levels of Government*

Value capture planning could be applied to any level of government, from local to state to federal.

9. *Integrating Value Capture into Other Planning Policies*

Value capture can be integrated into a range of other planning policies at different levels of government.

10. *Cooperative Planning Approach*

Value capture is a tool that lends itself to a cooperative planning approach (i.e. engaging government, developers, and the community).

7.6 Future Directions of Value Capture Within Urban Planning

The above results in applying the VCP model (7.2 and 7.3) and listed advantages (7.4 and 7.5) of VCP, it can be concluded that value capture can be a key planning tool. This would be especially so in the development and renewal of cities and regions. It was shown how VCP within future renewal precincts would provide more certainty in planning and development procedures as well as project finance. This includes implementing planning principles as illustrated in the case study precincts. These included equitable housing, public and open spaces and sustainable transport.

Future value capture programs can also embrace new design parameters and additional planning principles (utilising value capture tables). There are opportunities for all levels of government (local, state, and federal) to consider how best to utilise the value capture planning tool to meet a range of needs (i.e. arising from new development and renewal). The more value capture is used (cooperatively among stakeholders of government, developers, and the community) the more its effectiveness will help shape the future of planning cities and regions.

Reference

NSW Government (2018) Future transport 2056. Department of Transport, Sydney

Index

R. Rauscher, *Renewing Cities with Value Capture Planning*,
https://doi.org/10.1007/978-3-030-62958-8